Trussed Frames and Arches

Trussed Frames and Arches:

Schemes and Formulas

By

Mikhail Kirsanov

Cambridge
Scholars
Publishing

Trussed Frames and Arches: Schemes and Formulas

By Mikhail Kirsanov

This book first published 2020

Cambridge Scholars Publishing

Lady Stephenson Library, Newcastle upon Tyne, NE6 2PA, UK

British Library Cataloguing in Publication Data
A catalogue record for this book is available from the British Library

ISBN (10): 1-5275-5976-9
ISBN (13): 978-1-5275-5976-9

CONTENT

Introduction 1

Chapter 1. **Arches** 4
Truss 1.1 4
Truss 1.2 5
Truss 1.3 7
Truss 1.4 9
Truss 1.5 11
Truss 1.6 13
Truss 1.7 17
Truss 1.8 19
Truss 1.9 21
Truss 1.10 22
Truss 1.11 25
Truss 1.12 27
Truss 1.13 29
Truss 1.14 31
Truss 1.15 33
Truss 1.16 35
Truss 1.17 38

Chapter 2. **Frames** 41
Truss 2.1 41
Truss 2.2 43
Truss 2.3 45
Truss 2.4 47
Truss 2.5 49
Truss 2.6 51
Truss 2.7 53
Truss 2.8 56
Truss 2.9 59
Truss 2.10 62
Truss 2.11 63
Truss 2.12 66

Truss 2.13 68
Truss 2.14 70
Truss 2.15 72
Truss 2.16 74
Truss 2.17 76
Truss 2.18 78
Truss 2.19 80

Chapter 3. **Composite Frames and Arches** 84
Truss 3.1 84
Truss 3.2 85
Truss 3.3 87
Truss 3.4 89
Truss 3.5 91
Truss 3.6 93
Truss 3.7 95
Truss 3.8 97
Truss 3.9 99
Truss 3.10 102
Truss 3.11 105
Truss 3.12 107
Truss 3.13 109
Truss 3.14 111
Truss 3.15 113
Truss 3.16 115

Chapter 4. **Multi-span Frames and Brackets** 117
Truss 4.1 117
Truss 4.2 119
Truss 4.3 122
Truss 4.4 124
Truss 4.5 126
Truss 4.6 128
Truss 4.7 129
Truss 4.8 131
Truss 4.9 133

Truss 4.10 135

Chapter 5. **Cantilever trusses** 139
Truss 5.1 139
Truss 5.2 141
Truss 5.3 143
Truss 5.4 145
Truss 5.5 146
Truss 5.6 148
Truss 5.7 149
Truss 5.8 150
Truss 5.9 152
Truss 5.10 153

Chapter 6. **Program** 155
Trusses Schemes 161
References 173

Introduction

Nowadays, when numerical systems based on the finite element method have captured all the dominant heights in the field of structural analysis, accurate calculation formulas do not lose their significance. On the other hand, there are few such formulas and therefore the engineer has to turn to numerical methods. With this approach, one necessarily encounters standard shortcomings that are common for approximate calculations. First of all, there is a problem with calculations accuracy. In a simple case, when the design is simple and the number of its elements, for example, rods, is limited, this inaccuracy is not that significant. However, dealing with very large-scale structures containing thousands of elements is a real issue. Take as an example the overlapping of stadiums or large-span bridges (Fig. 1 – 3). Doing calculations in such cases is laborious, and no one can guarantee the accuracy of them. The well-known problem, so called the «curse of dimension», is a guaranteed loss of accuracy when solving a system of linear equations with the number of equations increased.

Fig. 1. Royal Albert Bridge, river Tamar, United Kingdom

For large order of the system, the solution for individual unknowns may have an error of 100% or more. In such cases simple and reliable formulas, containing as parameters not only the dimensions of the structure, but also the number of elements could be very useful. The number of elements here does not affect accuracy. Such formulas can be derived by induction for regular systems having periodicity elements in their structure. Experience shows that such formulas can be derived by induction for regular systems having periodicity elements in their structure. For trusses, the periodicity element may be a panel containing a number of rods. The periodicity element for trusses could be a panel containing a number of rods.

Fig. 2. Story Bridge, Brisbane, Queensland, Australia

This handbook provides a very practical and simple approach for overcoming these issues. There are 72 schemes of planar statically determinate trusses with formulas for the dependence of deflection and forces in some (usually critical) rods on the number of panels.

Fig. 3. Andreyevsky Bridge, Moscow, Russia

For each schema formulas are given for two cases: distributed or concentrated load. For the structures that contain a movable support, there are additional formulas for the magnitude of its horizontal displacement. This handbook consists of six chapters. Trusses schemes are conventionally divided into certain types in the first five chapters. The division by trusses types is not straightforward. For example, some trusses may belong to both arches and frames. Chapter 5 shows console calculations. All the schemes in the handbook are original, except the simplest ones. The sixth chapter describes the program written in Maple [76] language. In addition to the problem of allocating time for work, by breaking away from teaching, the author had two difficulties while he was writing this handbook.

The first and main problem was to come up with an original scheme of a statically determinate regular truss in a way that is kinematically unchanged, and the formula for its calculation would not be very cumbersome.

Publishing voluminous solutions containing half page trigonometric functions, is not a good idea due to small practical value. It is not a practical approach to use a formula the perfection of which is hard to check due to its size and complication. However, it is a good point here to talk about the second challenge, which is related to the reliability of the obtained solutions.

How do I guarantee the reliability of the derived formulas? The following technique was found, previously used by the author in his first handbook [28]. Firstly, the conclusion is duplicated and in order to avoid elementary typos, it is checked by typing from a sheet of a finished manuscript in numerical form. Secondly, all formulas for each truss are placed in separate files in text format. These files are available for free download (http://vuz.exponenta.ru/Trusstxt2020.zip). Thus, one has an opportunity to avoid typing large formulas from the book sheet. Just copy the formula from the file and place it into the program that performs calculations. I think, the best approach is to use Pascal and Maple based languages for such programs.

The handbook is a continuation of the first handbook [28]. The author hopes that the formulas could be useful to engineers in their practice, to students — for training purposes. The initial idea of generating formulas for calculating regular systems (not necessarily trusses) could be helpful for researchers in their scientific activities.

Please do not hesitate to contact the author via following email: c216@ya.ru.

Chapter 1

Arches

Truss 1.1

A truss with a height of $(m+1)h$, $h = a\sqrt{3}/2$ consists of $8n + 8m + 3$ rods of length a (Fig. 4).

1.1.1. Concentrated force in the mid-span

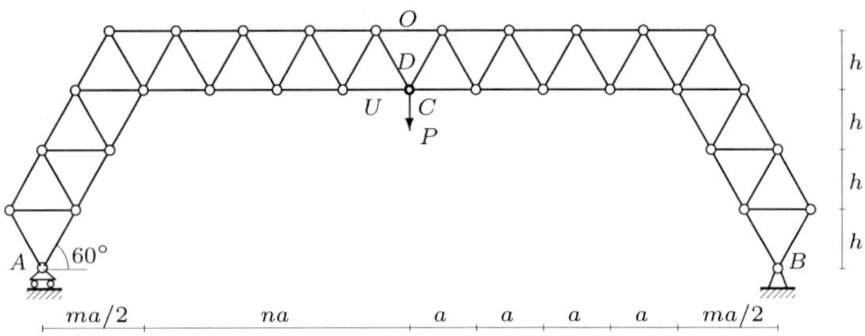

Fig. 4. Truss, $m = 3$, $n = 4$

The deflection (vertical displacement of the middle node C in the lower zone) has the form

$$\Delta = Pa(4m^3 + 3(1+4n)m^2 + 2(13+12n^2)m+$$
$$+2n(25+8n^2))/(36EF).$$

Support A offset:

$$\delta_A = Pa\sqrt{3}\,(4m^3+3(3+4n)m^2+2(6n^2+3n+4)m+6n^2)/(18EF).$$

Forces:

$$O = -P\sqrt{3}\,(2n+m)/6,\ D = P\sqrt{3}/3,\ U = P\sqrt{3}\,(2n-1+m)/6.$$

Supports reactions: $Y_A = Y_B = P/2$, $X_B = 0$.

1.1.2. Load on the upper belt

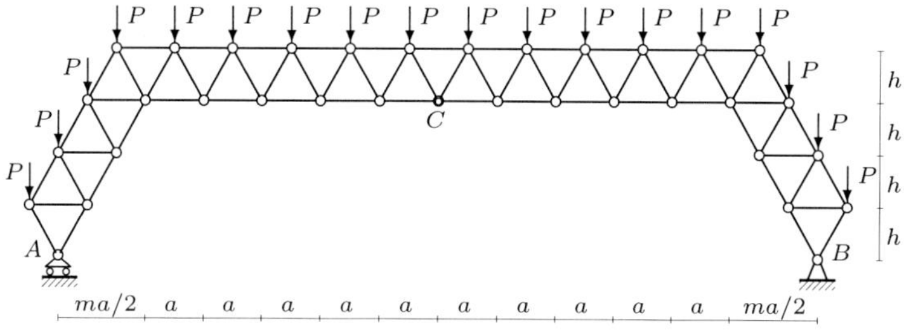

Fig. 5. Truss, $m = 3$, $n = 5$

The deflection (vertical displacement of the middle node C in the lower zone) has the form

$$\Delta = Pa(5m^4 + (20n - 1)m^3 + (36n^2 - 6n + 10)m^2 + \\ +(40n^3 - 12n^2 + 30n + 16)m + 4n^2(7 + 5n^2))/(36EF).$$

Support A offset:

$$\delta_A = Pa\sqrt{3}\,(5m^4 + (7 + 20n)m^3 + 2(12n^2 + 6n - 7)m^2 + \\ +4(n - 1)(4n^2 + 7n + 4)m + 8n(n^2 - 1))/18.$$

Forces:

$$O = -P\sqrt{3}\,(m^2 + (2n - 1)m + 2(n^2 - 1))/6,\ D = 0,\ U = -O.$$

Supports reactions: $Y_A = Y_B = (n + m)P$, $X_B = 0$.

Truss 1.2

A truss with a height of $h + 2b$ (Fig. 6) containing $2n + 4$ panels, consists of $8n + 17$ rods. The total length of the rods in the truss is

$$L_{sum} = 2(2a + c)(n + 1) + 2d + g + 2h(n + 4),$$

where $c = \sqrt{a^2 + h^2}$, $d = \sqrt{4a^2 + h^2}$, $g = \sqrt{4a^2 + 9h^2}$, $b = h/2$.

The deflection (vertical displacement of the middle node C in the lower zone) has the form

$$\Delta = P(C_1a^3 + C_2c^3 + C_3h^3 + C_4(d^3 + g^3))/(6h^2EF). \qquad (1.1)$$

Support A offset:

$$\delta_A = P(A_1a^3 + A_2c^3 + A_3h^3 + A_4d^3 + A_5g^3)/(9ahEF). \qquad (1.2)$$

1.2.1. Concentrated force in the mid-span

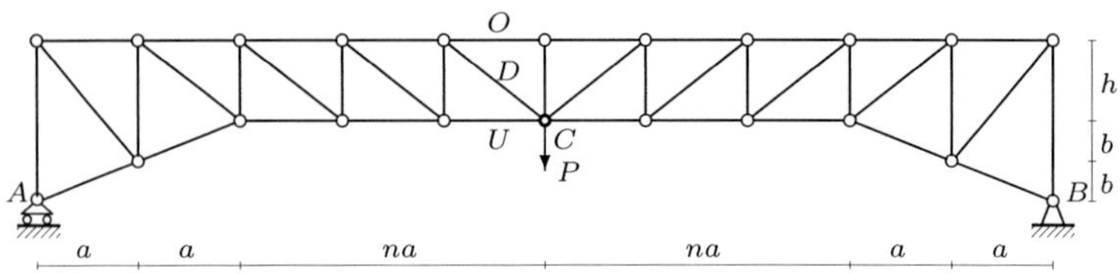

Fig. 6. Truss, $n = 3$

Deflection. The coefficients in (1.1):

$C_1 = (6n^3 + 36n^2 + 75n + 40)/3$, $C_2 = (9n + 16)/3$,

$C_3 = 14 + 3n$, $C_4 = 1/6$.

Support offset. The coefficients in (1.2):

$A_1 = (27n^2 + 99n + 40)/2$, $A_2 = 8$, $A_3 = 21$, $A_4 = 1$, $A_5 = 1/4$.

Forces:

$O = -Pa(n + 2)/(2h)$, $D = Pc/(2h)$, $U = Pa(n + 1)/(2h)$.

Supports reactions: $Y_A = Y_B = P/2$, $X_B = 0$.

1.2.2. Load on the upper belt

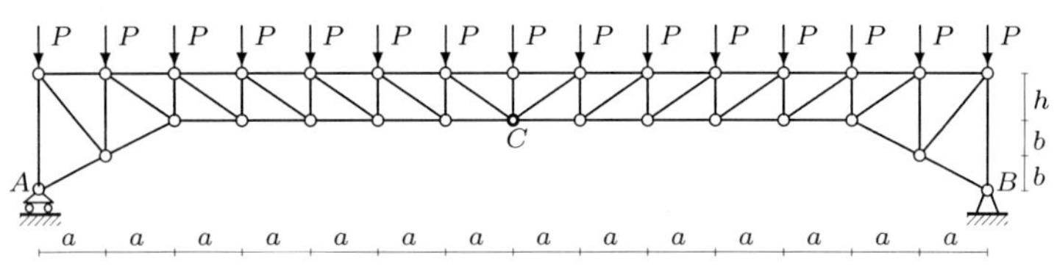

Fig. 7. Truss, $n = 5$

Deflection. The coefficients in (1.1):

$C_1 = (15n^4 + 120n^3 + 363n^2 + 436n + 168)/6$,

$C_2 = (9n^2 + 32n + 24)/3$, $C_3 = 3n^2 + 34n + 54$,

$C_4 = (2n + 3)/6$.

Support offset. The coefficients in (1.2):

$A_1 = (36n^3 + 207n^2 + 287n + 84)/2$,

$A_2 = 16n + 12$, $A_3 = 42n + 81$,

$A_4 = 2n + 3$, $A_5 = (2n + 3)/4$.

Forces:

$O = -Pa(n + 2)^2/(2h)$, $D = Pc/(2h)$,

$U = Pa(n + 1)(n + 3)/(2h)$.

Supports reactions:

$Y_A = Y_B = (2n + 5)P/2$, $X_B = 0$.

Truss 1.3

A truss with a height of $(2n + 1)h$ (Fig. 8) containing $2n$ panels, consists of $8n - 1$ rods [6].

The deflection (vertical displacement of the middle node C in the lower zone) has the form

$$\Delta = P(C_1a^3 + C_2c^3 + C_3h^3)/(h^2EF), \tag{1.3}$$

where $c = \sqrt{a^2 + h^2}$.

The total length of the rods in the truss is

$$L_{sum} = 4(3c + h)n + 2a(1 + 2n) - 2c.$$

Support A offset:

$$\delta_A = P(A_1a^3 + A_2c^3 + A_3h^3)/(ahEF). \tag{1.4}$$

1.3.1. Concentrated force in the mid-span

Deflection. The coefficients in (1.3):

$C_1 = n(2n + 1)$, $C_2 = n(2n^2 + 1)/3$, $C_3 = n$.

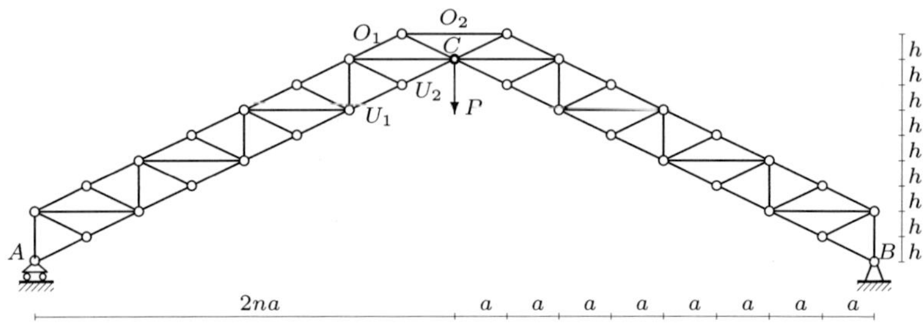

Fig. 8. Truss, $n = 4$

Support offset. The coefficients in (1.4):

$A_1 = 2n(2n + 1)$, $A_2 = n(4n^2 + 3n - 1)/3$, $A_3 = 2n$.

Forces:

$O_1 = -Pcn/(2h)$, $O_2 = -Pan/h$,

$U_1 = U_2 = Pc(n - 1)/(2h)$, $V = -P/2$.

Supports reactions: $Y_A = Y_B = P/2$, $X_B = 0$.

1.3.2. Load on the upper belt

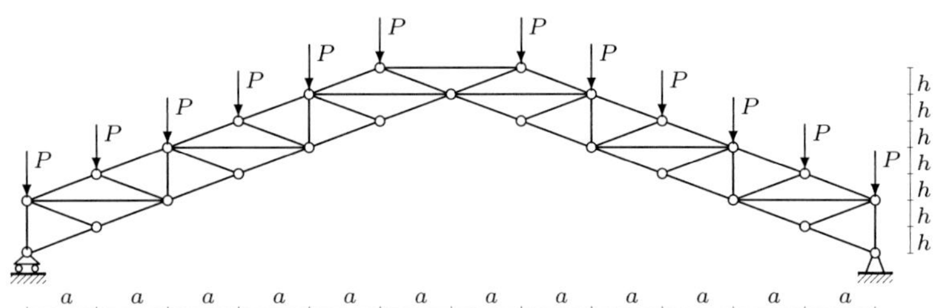

Fig. 9. Truss, $n = 3$

Deflection. The coefficients in (1.3):

$C_1 = 4n^3$, $C_2 = n(20n^3 - 8n^2 + 7n - 7)/12$, $C_3 = 2n(n + 1)$.

Support offset. The coefficients in (1.4):

$A_1 = 8n^3$, $A_2 = n(20n^3 + 8n^2 - 11n - 5)/6$, $A_3 = 4n(1 + n)$.

Forces:

$O_1 = -Pc(2n^2 - n + 1)/(2h)$, $O_2 = -Pan(2n - 1)/h$,

$U_1 = U_2 = Pc(2n^2 - n - 1)/(2h)$, $V = -2P$.

Supports reactions: $Y_A = Y_B = 2nP$, $X_B = 0$.

Truss 1.4

A truss with a height of $2(n + 1)h$ (Fig. 8) containing $2n$ panels, consists of $8n + 1$ rods. The total length of the rods in the truss is

$$L_{sum} = 4(a + 3c + h)n + 2h,$$

where $c = \sqrt{a^2 + h^2}$.

The deflection (vertical displacement of the middle node C in the lower zone) has the form

$$\Delta = P(C_1 a^3 + C_2 c^3 + C_3 h^3)/(h^2 EF). \qquad (1.5)$$

Support A offset:

$$\delta_A = P(A_1 a^3 + A_2 c^3 + A_3 h^3)/(ahEF). \qquad (1.6)$$

1.4.1. Concentrated force in the mid-span

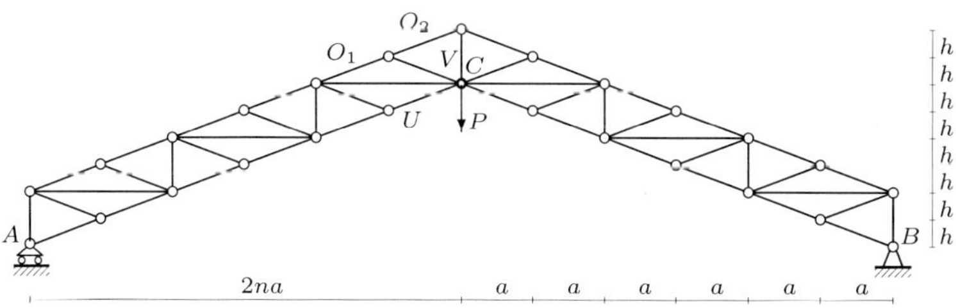

Fig. 10. Truss, $n = 3$

Deflection. The coefficients in (1.5):

$C_1 = n$, $C_2 = n(1 + 2n^2)/3$, $C_3 = n(1 + 2n)$.

Support offset. The coefficients in (1.6):

$A_1 = 2n$, $A_2 = n(4n^2 + 3n - 1)/3$, $A_3 = 2n(1 + 2n)$.

Forces:

$O_1 = O_2 = -Pcn/(2h)$, $U = Pc(n - 1)/(2h)$, $V = Pn$.

Supports reactions: $Y_A = Y_B = P/2$, $X_B = 0$.

1.4.2. Load on the upper belt

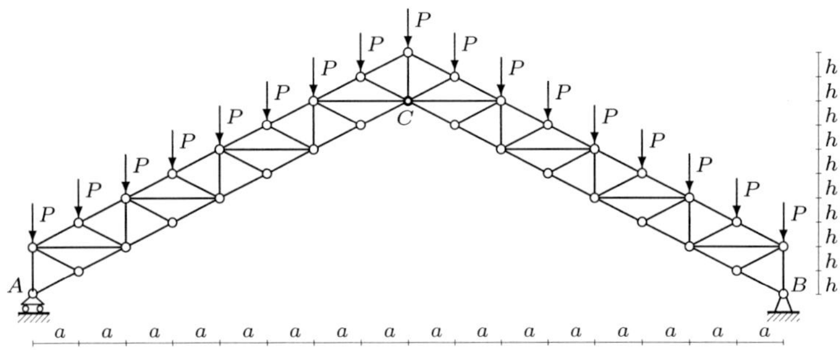

Fig. 11. Truss, $n = 4$

Deflection. The coefficients in (1.5):

$C_1 = n(2n + 1)$, $C_2 = n(20n^3 + 7n + 3)/12$,

$C_3 = n(4n^2 + 2n + 1)$.

Support offset. The coefficients in (1.6):

$A_1 = 2n(1 + 2n)$, $A_2 = n(20n^3 + 16n^2 - 5n - 1)/6$,

$A_3 = 2n(4n^2 + 2n + 1)$.

Forces:

$O_1 = -Pc(1 + 2n^2)/(2h)$, $O_2 = -Pcn^2/h$,

$U = Pc(n^2 - 1)/h$, $V = (2n^2 - 1)P$.

Supports reactions: $Y_A = Y_B = (4n + 1)P/2$, $X_B = 0$.

Truss 1.5

A truss with a height of $nf+(m+1)h$ (Fig. 12) containing $2(n+$ $+m)$ panels, consists of $8(n+m)+1$ rods. The total length of the rods in the truss is

$$L_{sum} = 2ma + 4mc + 3dn + (2m + 2n + 1)h,$$

where $c = \sqrt{a^2+h^2}$, $d = \sqrt{4a^2+h^2}$, $f = h/2$.

The deflection (vertical displacement of the middle node C in the lower zone) has the form

$$\Delta = P(C_1a^3 + C_2c^3 + C_3d^3 + C_4h^3)/(16h^2EF). \qquad (1.7)$$

Support A offset has the form

$$\delta_A = P(A_1a^3 + A_2c^3 + A_3d^3 + A_4h^3)/(8ahEF). \qquad (1.8)$$

1.5.1. Concentrated force in the mid-span

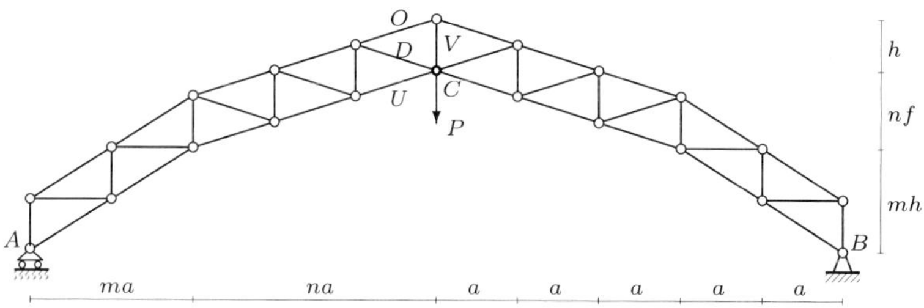

Fig. 12. Truss, $n - 3$, $m = 2$

Deflection. The coefficients in (1.7):

$C_1 = 8m$, $C_2 = 8m(1+2m^2)/3$,

$C_3 = 2(n^3 + 3n^2m + (2+3m^2)n)/3$,

$C_4 = 2(2n^2 + 4(m+1)n + 3m^2)$.

Support offset. The coefficients in (1.8):

$A_1 = 8m$, $A_2 = 4m(4m^2+3m-1)/3$,

$A_3 = (2n^3 + 3(3m+1)n^2 + (12m^2 + 6m + 1)n)/6,$

$A_4 = 2(n^2 + (3m+2)n + m + 3m^2).$

Forces:

$O = -Pd(n+m)/(4h),\ D = Pd/(4h),$

$U = Pd(n+m-1)/(4h),\ V = P(n+m)/2.$

Supports reactions: $Y_A = Y_B = P/2,\ X_B = 0.$

1.5.2. Load on the upper belt

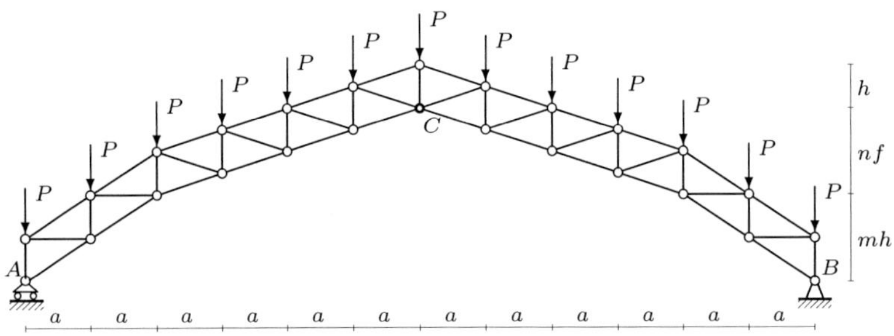

Fig. 13. Truss, $n = 4$, $m = 2$

Deflection. The coefficients in (1.7):

$C_1 = 8m(2n+m),\ C_2 = 4m(4(1+2m^2)n + m(1+5m^2))/3,$

$C_3 = (5n^4 + 20n^3m + (7+30m^2)n^2 + 2m(6m^2-1)n)/6,$

$C_4 = 2(2n^3 + 2(2+3m)n^2 + 4(2m^2+1)n + 3m^3 + 2m^2 + 2m).$

Support offset. The coefficients in (1.8):

$A_1 = 8m(2n+m),$

$A_2 = 4m(m+1)(2(4m-1)n + 5m^2 - m - 1)/3,$

$A_3 = (5n^4 + 4(2+7m)n^3 + (1+24m+54m^2)n^2+$

$+2(2m+1)(6m^2-1)n)/12,$

$A_4 = 2(n^3 + (4m+2)n^2 + (7m^2+2m+2)n + m(3m^2+3m+2)).$

Forces:

$O = -Pd(n+m)^2/(4h)$, $D = Pd/(4h)$,

$U = Pd((n+m)^2 - 1)/(4h)$, $V = P((n+m)^2 - 2)/2$.

Supports reactions: $Y_A = Y_B = (2n + 2m + 1)P/2$, $X_B = 0$.

Truss 1.6

A truss with a height of $(m+1)h$ (Fig. 14) containing $2(n+m+2)$ panels, consists of $8(m+n)+7$ rods. The total length of the rods in the truss is

$$L_{sum} = 4(n+1)a + 2(2m+1)c + 2(m+n+1)d + 2(m+n)h,$$

where $c = \sqrt{a^2+h^2}$, $d = \sqrt{4a^2+h^2}$.

The deflection (vertical displacement of the middle node C in the upper zone) has the form

$$\Delta = P(C_1a^3 + C_2c^3 + C_3d^3 + C_4h^3)/(2h^2EF). \tag{1.9}$$

Support A offset has the form

$$\delta_A = P(A_1a^3 + A_2c^3 + A_3d^3 + A_4h^3)/(ahEF). \tag{1.10}$$

1.6.1. Concentrated force in the mid-span

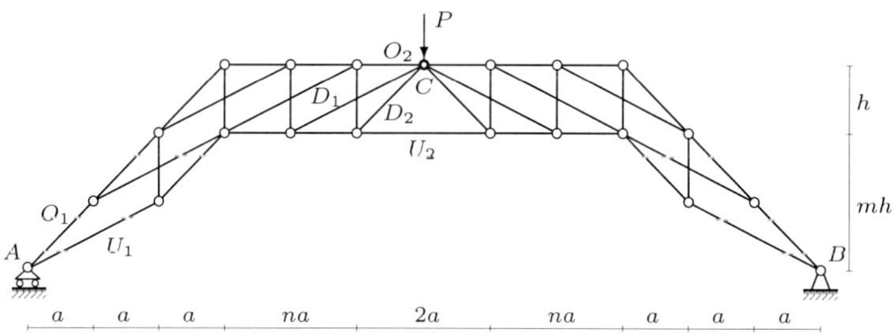

Fig. 14. Truss, $n = 2$, $m = 2$

Deflection. The coefficients in (1.9):

$$C_1 = (4n^3 + 12(m+2)n^2 + 4(3(-1)^m + 17 + 12m + 3m^2)n+$$
$$+12(-1)^m - 3(-1)^{n+m} + 12m^2 + 42m+$$
$$+45 + 6(-1)^m m - 6(-1)^n)/6,$$

$$C_2 = (12m^2 + 26m + 27 + 4m^3 + 12(-1)^m + 3(-1)^{n+m} + \\ +6(-1)^m m + 6(-1)^n)/6,$$

$$C_3 = (4(2(-1)^m + 3)n + 4m + 10 - 2(-1)^{n+m} - \\ -4(-1)^n + 4(-1)^m)/8,$$

$$C_4 = (2(2(-1)^m + 3)n + 4(-1)^m m + 6(-1)^m + \\ +4m^2 + 14m + 7 + (-1)^{n+m} + 2(-1)^n)/4,$$

Support offset. The coefficients in (1.10):

$$A_1 = (2(2m + 1)n^2 + 2(4m^2 + 10m + 3(-1)^m + 9)n + 20m + 15 + \\ +6(-1)^m + 8m^2 - 3(-1)^n + 4(-1)^m m - 2(-1)^{n+m})/4,$$

$$A_2 = (8m^3 + 18m^2 + 4(3(-1)^m + 7)m + \\ +3(-1)^n + 24 + 18(-1)^m + 3(-1)^{n+m})/12,$$

$$A_3 = (6(1 + (-1)^m)n + 4m + 7 - (-1)^n - (-1)^{n+m} + 3(-1)^m)/8,$$

$$A_4 = (8m^2 + 4(2(-1)^m + 5)m + 5 + 9(-1)^m + \\ +(-1)^{n+m} + 6(-1)^m n + (-1)^n + 6n)/8.$$

Forces:

$O_1 = -Pc/h$, $O_2 = -Pa(2n + (-1)^n(2 + (-1)^m) + 2m + 1)/(4h)$,

$D_1 = Pd((-1)^n(2 + (-1)^m) - 1)/(4h)$,

$D_2 = Pc((-1)^n(2 + (-1)^m) + 1)/(4h)$,

$U_1 = Pd/(2h)$, $U_2 = Pa(n + 2 + m)/(2h)$.

Supports reactions:

$Y_A = Y_B = P/2$, $X_B = 0$.

1.6.2. Load on the upper belt

Deflection. The coefficients in (1.9):

$$C_1 = (10n^4 + 40(2 + m)n^3 + 4(15m^2 + 60m + 77 + 12(-1)^m)n^2 + \\ +2(12m^3 + 96m^2 + 8(29 + 3(-1)^m)m + 221 - 6(-1)^n - \\ -3(-1)^{n+m} + 60(-1)^m)n + 24m^3 + 6(23 + 3(-1)^m)m^2 +$$

$$+6(45-(-1)^{n+m}+12(-1)^m)m+198-18(-1)^n-15(-1)^{n+m}+$$
$$+75(-1)^m)/12,$$

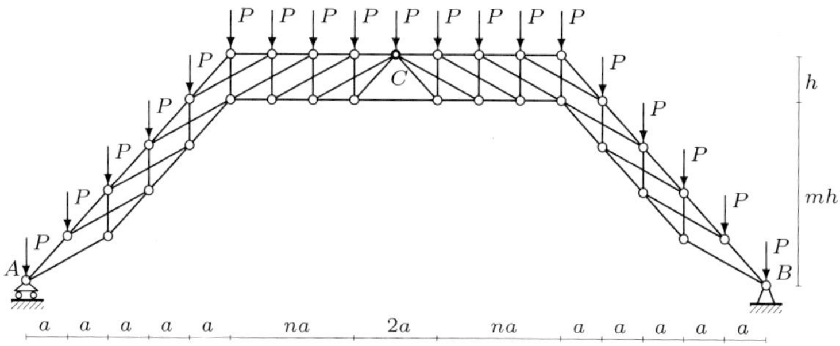

Fig. 15. Truss, $n = 3$, $m = 4$

$$C_2 = (10m^4 + 8(2n+7)m^3 + 2(67 + 24n + 9(-1)^m)m^2+$$
$$+2(4(3(-1)^m + 13)n + 101 + 3(-1)^{n+m} + 36(-1)^m)m+$$
$$+6(17 + 2(-1)^n + (-1)^{m+n} + 8(-1)^m)n + 141 + 75(-1)^m+$$
$$+12(-1)^{n+m} + 12(-1)^n)/12,$$
$$C_3 = (2(11 + 8(-1)^m)n^2 + 2(2(2(-1)^m + 3)m + 23 - 2(-1)^n+$$
$$+16(-1)^m-(-1)^{n+m})n+6m^2+2(11-(-1)^{n+m}+2(-1)^m)m+$$
$$+10(-1)^m + 25 - 5(-1)^{n+m} - 6(-1)^n)/8,$$
$$C_4 = (8m^3 + 2(8n + 25 + 6(-1)^m)m^2 + 2(41 + 12(-1)^m n+$$
$$+22(-1)^m + (-1)^{n+m} + 30n)m + 2(11 + 8(-1)^m)n^2+$$
$$+2(25 + (-1)^{n+m} + 2(-1)^n + 24(-1)^m)n+$$
$$+21 + 2(-1)^n + 38(-1)^m + 3(-1)^{n+m})/8.$$

Case $m = n$:

$$C_1 = (134n^4 + 536n^3 + 2(455 + 57(-1)^n)n^2 + 20(35 + 9(-1)^n)n+$$
$$+183 + 57(-1)^n)/12,$$
$$C_2 = (26n^4 + 104n^3 + 14(17 + 3(-1)^n)n^2 + 4(79 + 33(-1)^n)n+$$
$$+153 + 87(-1)^n)/12,$$

$$C_3 = (2(5+3(-1)^n)n^2 + 8(2+(-1)^n)n + (-1)^n + 5)/2,$$

$$C_4 = (6n^3 + (33+13(-1)^n)n^2 + (24(-1)^n+34)n + 6 + 10(-1)^n)/2.$$

Support offset. The coefficients in (1.10):

$$A_1 = (8(1+2m)n^3 + 12(4m^2+10m+8+3(-1)^m)n^2 +$$
$$+(36(-1)^m m + 190 - 12(-1)^n + 272m - 6(-1)^{n+m} +$$
$$+84(-1)^m + 156m^2 + 24m^3)n + 99 + 174m + 18(-1)^m m^2 +$$
$$+60(-1)^m m - 15(-1)^n + 114m^2 - 15(-1)^{n+m} + 24m^3 +$$
$$+51(-1)^m - 6m(-1)^{n+m})/12,$$

$$A_2 = (4(12+9(-1)^m + 4m^3 + 6(-1)^m m + 14m + 9m^2)n + 69 +$$
$$+108m + 18(-1)^m m^2 + 60(-1)^m m + 10m^4 + 48m^3 + 86m^2 +$$
$$+51(-1)^m)/12,$$

$$A_3 = (12((-1)^m+1)n^2 + 2(6m + 2(-1)^m m + 11(-1)^m + 15)n +$$
$$+22m + 20 + 2(-1)^m m - (-1)^{n+m} + 6m^2 + 6(-1)^m - (-1)^n)/8,$$

$$A_4 = (12((-1)^m+1)n^2 + 2(8m^2 + 2(5(-1)^m + 11)m + 17(-1)^m +$$
$$+13)n + 8 + 50m + 8m^3 - (-1)^n + 38(-1)^m m -$$
$$-(-1)^{n+m} + 26(-1)^m + 6(2(-1)^m + 7)m^2)/8.$$

Case $m = n$:

$$A_1 = (44n^4 + 154n^3 + (45(-1)^n + 241)n^2 +$$
$$+2(88 + 33(-1)^n)n + 18(-1)^n + 42)/6,$$

$$A_2 = (26n^4 + 84n^3 + 2(21(-1)^n + 71)n^2 + 12(13 + 8(-1)^n)n +$$
$$+69 + 51(-1)^n)/12,$$

$$A_3 = (2(15 + 8(-1)^n)n^2 + 4(6(-1)^n + 13)n + 19 + 5(-1)^n)/8,$$

$$A_4 = (24n^3 + (98 + 44(-1)^n)n^2 + 4(18(-1)^n + 19)n +$$
$$+7 + 25(-1)^n)/8.$$

Forces:

$$O_1 = -Pc(2m + 2n + 3)/h,$$

$$O_2 = -Pa(m^2 + 4m + 2 + 2nm + 4n + (-1)^{n+m}(m + n + 2)+$$
$$+2(-1)^n(1 + n) + n^2)/(2h),$$
$$D_1 = Pd((-1)^{n+m}(n + m + 2) + 2(-1)^n(1 + n) - 1)/(2h),$$
$$D_2 = -Pc((-1)^{n+m}(n + m + 2) + 2(-1)^n(1 + n))/(2h),$$
$$U_1 = Pd(2m + 2n + 3)/(2h),$$
$$U_2 = Pa(m^2 + 4m + 4 + 2nm + 4n + n^2)/(2h).$$

Supports reactions:

$$Y_A = Y_B = P(2n + 2m + 5)/2,\ X_B = 0.$$

Truss 1.7

A truss with a height of $nf + h$ (Fig. 16) containing $2n$ panels, consists of $8n + 1$ rods. The total length of the rods in the truss is

$$L_{sum} = 2(2c + h + d)n + h,$$

where $c = \sqrt{a^2 + f^2}$, $d = \sqrt{a^2 + (h - f)^2}$.

1.7.1. Concentrated force in the mid-span

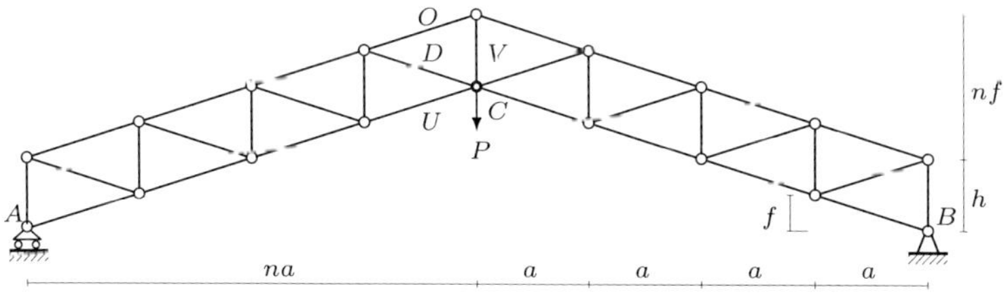

Fig. 16. Truss, $n = 4$

The deflection (vertical displacement of the middle node C in the lower zone) has the form

$$\Delta = Pn(2c^3n^2 + 6f^2hn + c^3 + 3d^3 + 3h^3)/(6h^2EF).$$

Support A offset has the form

$$\delta_A = Pn(4c^3fn^2 + 3h(4f^3 + c^3)n + 2c^3f - 3c^3h+$$

$+6d^3f+6fh^3)/(6h^2aEF)$.

Forces:

$O=-Pcn/(2h)$, $D=Pd/(2h)$,

$U=Pc(n-1)/(2h)$, $V=Pfn/h$.

Supports reactions:

$Y_A=Y_B=P/2$, $X_B=0$.

1.7.2. Load on the upper belt

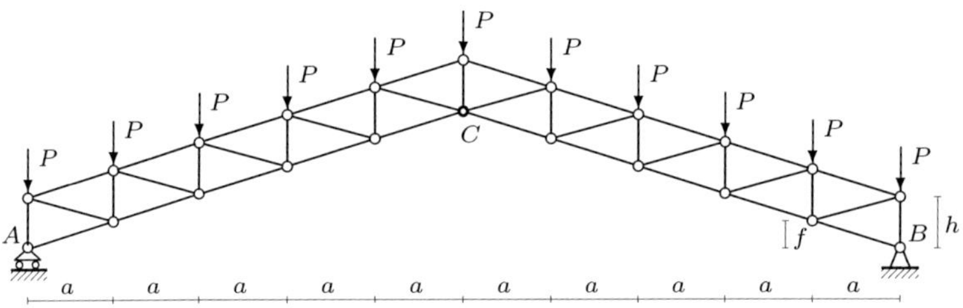

Fig. 17. Truss, $n=5$

The deflection (vertical displacement of the middle node C in the lower zone) has the form

$$\Delta=Pn(5c^3n^3+12f^2hn^2+(c^3+6h^3+6d^3)n+12h^2(h-f))/(12h^2EF).$$

Support A offset has the form

$$\delta_A=Pn(5c^3fn^3+4h(c^3+3f^3)n^2+(6d^3f+c^3f+6fh^3-3c^3h)n-$$
$$-h(c^3-12hf(h-f)))/(6ah^2EF).$$

Forces:

$O=-Pcn^2/(2h)$, $D=Pd/(2h)$,

$U=Pc(n^2-1)/(2h)$, $V=P(fn^2/h-1)$.

Supports reactions:

$Y_A=Y_B=P(2n+1)/2$,

$X_B=0$.

Truss 1.8

A truss with a height of $(n+1)h$ (Fig. 18) containing $2n+2$ panels, consists of $8n+5$ rods. The total length of the rods in the truss is

$$L_{sum} = 2(n+1)a + 2(2n+1)c + (2n+1)h,$$

where $c = \sqrt{a^2+h^2}$.

The deflection (vertical displacement of the middle node C in the lower zone) has the form

$$\Delta = P(C_1a^3 + C_2c^3 + C_3h^3)/(h^2EF). \tag{1.11}$$

Support A offset:

$$\delta_A = P(A_1a^3 + A_2c^3 + A_3h^3)/(ahEF). \tag{1.12}$$

1.8.1. Concentrated force in the mid-span

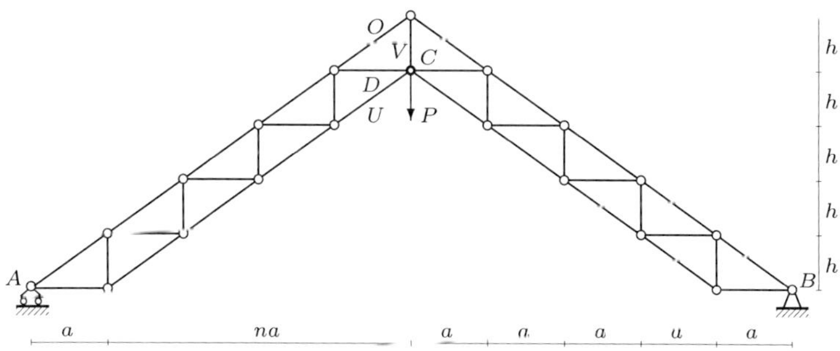

Fig. 18. Truss, $n = 4$

Deflection. The coefficients in (1.11):

$C_1 = (n+1)/2,$

$C_2 = (2n^3 + 6n^2 + 7n + 3)/6,$

$C_3 = (2n^2 + 5n + 2)/2.$

Support offset. The coefficients in (1.12):

$A_1 = n+1,\ A_2 = n(4n^2 + 9n + 5)/6,\ A_3 = n(2n+3).$

Forces:

$O = -Pc(n+1)/(2h)$, $D = Pa/(2h)$,

$U = Pcn/(2h)$, $V = P(n+1)$.

Supports reactions:

$Y_A = Y_B = P/2$, $X_B = 0$.

1.8.2. Load on the upper belt

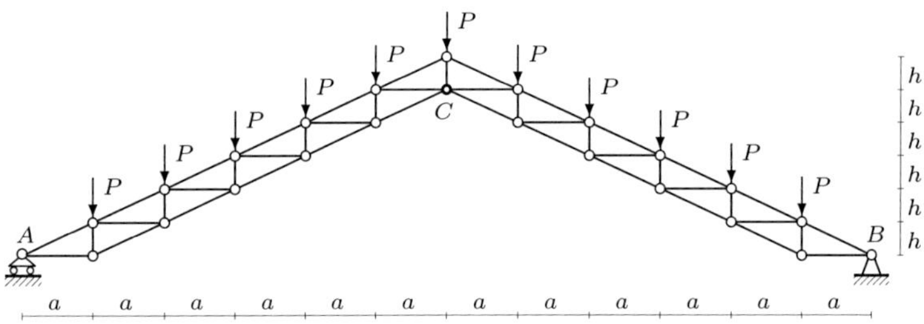

Fig. 19. Truss, $n = 5$

Deflection. The coefficients in (1.11):

$C_1 = (n+1)^2/2$,

$C_2 = (5n^4 + 20n^3 + 31n^2 + 22n + 6)/12$,

$C_3 = n(2n^2 + 7n + 6)/2$.

Support offset. The coefficients in (1.12):

$A_1 = (n+1)^2$, $A_2 = n(5n^3 + 16n^2 + 16n + 5)/6$,

$A_3 = n(2n^2 + 5n + 2)$.

Forces:

$O = -Pc(n+1)^2/(2h)$, $D = Pa/(2h)$,

$U = Pcn(n+2)/(2h)$, $V = Pn(n+2)$.

Supports reactions:

$Y_A = Y_B = P(2n+1)/2$, $X_B = 0$.

Truss 1.9

A truss with a height of $(n+1)h$ (Fig. 20) containing $2n$ panels, consists of $8n+1$ rods. The total length of the rods in the truss is

$$L_{sum} = 2(a+2c+h)n+h,$$

where $c = \sqrt{a^2+h^2}$.

The deflection (vertical displacement of the middle node C in the upper zone) has the form

$$\Delta = P(C_1a^3 + C_2c^3 + C_3h^3)/(h^2EF). \tag{1.13}$$

Support A offset has the form

$$\delta_A = P(A_1a^3 + A_2c^3 + A_3h^3)/(ahEF). \tag{1.14}$$

1.9.1. Concentrated force in the mid-span

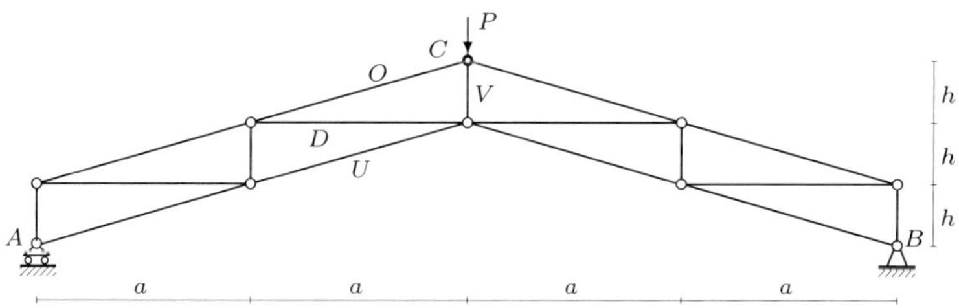

Fig. 20. Truss, $n-2$

Deflection. The coefficients in (1.13):

$C_1 = n/2$, $C_2 = n(2n^2+1)/6$, $C_3 = (2n^2-3n+2)/2$.

Support offset. The coefficients in (1.14):

$A_1 = n$, $A_2 = n(4n^2+3n-1)/6$, $A_3 = n(2n-1)$.

Forces:

$O = -Pcn/(2h)$, $D = Pa/(2h)$,

$U = Pc(n-1)/(2h)$, $V = P(n-1)$.

Supports reactions: $Y_A = Y_B = P/2$, $X_B = 0$.

1.9.2. Load on the upper belt

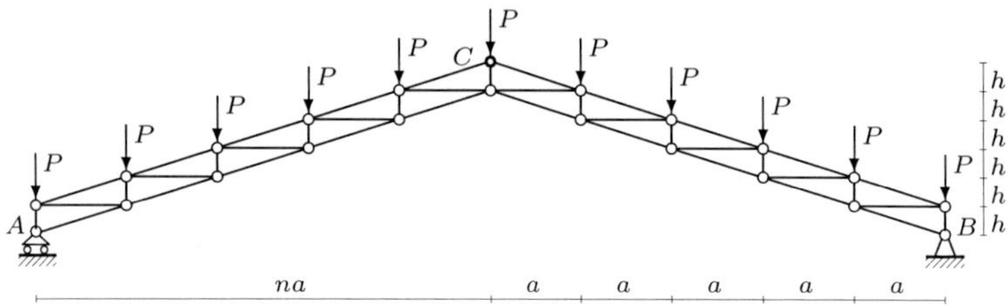

Fig. 21. Truss, $n = 5$

Deflection. The coefficients in (1.13):

$C_1 = n^2/2$, $C_2 = n^2(1 + 5n^2)/12$, $C_3 = (2n^3 - n^2 + 2)/2$.

Support offset. The coefficients in (1.14):

$A_1 = n^2$, $A_2 = n(5n^3 + 4n^2 - 2n - 1)/6$, $A_3 = n^2(2n + 1)$.

Forces:

$O = -Pcn^2/(2h)$, $D = Pa/(2h)$,

$U = Pc(n^2 - 1)/(2h)$, $V = P(n^2 - 1)$.

Supports reactions: $Y_A = Y_B = P(2n + 1)/2$, $X_B = 0$.

Truss 1.10

A truss with a height of $(n + m)h$ (Fig. 22) containing $2n + 2m$ panels, consists of $8n + 8m + 5$ rods. The total length of the rods in the truss is

$$L_{sum} = (2m + 2n + 3)a + 3mc + 4gn + (1 + 2n)h,$$

where $c = \sqrt{a^2 + 4h^2}$, $g = \sqrt{a^2 + h^2}$.

The case $d = a$, $b = a/2$ is considered.

The deflection (vertical displacement of the middle node C in the upper zone) has the form

$$\Delta = P(C_1a^3 + C_2c^3 + C_3g^3 + C_4h^3)/(h^2EF). \tag{1.15}$$

Support A offset:

$$\delta_A = P(A_1 a^3 + A_2 c^3 + A_3 g^3 + A_4 h^3)/(ahEF). \tag{1.16}$$

1.10.1. Concentrated force in the mid-span

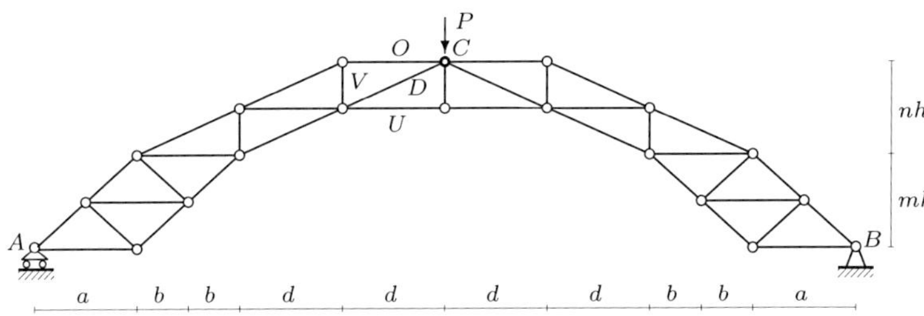

Fig. 22. Truss, $n = m = 2$

Deflection. The coefficients in (1.15):

$C_1 = (24n^2 + 8(4 + 3m)n + 7m^2 + 20m + 8)/32,$

$C_2 = m(m^2 + 3m + 5)/96,$

$C_3 = (8n^3 + 12n^2 m + 2(3m^2 + 2)n - 3(m^2 - 4))/24,$

$C_4 = (4n^2 + 4(1 + m)n + m^2 - 4)/8.$

Support offset. The coefficients in (1.16):

$A_1 = (12n^2 + 2(9m + 2)n + m(7m + 10))/8,$

$A_2 = m(m + 1)(2m + 1)/48,$

$A_3 = (4n^3 + 3(3m - 1)n^2 + (6m^2 - 3m - 1)n - 3m^2)/6,$

$A_4 = (2n^2 + 3nm + (m + 1)(m - 2))/2.$

Forces:

$O = -Pa(m + 2n)/(4h),\ D = -Pg/(2h),$

$U = Pa(m + 2n + 2)/(4h),$

$V = P(m + 2n)/4.$

Supports reactions:

$Y_A = Y_B = P/2,\ X_B = 0.$

1.10.2. Load on the upper belt

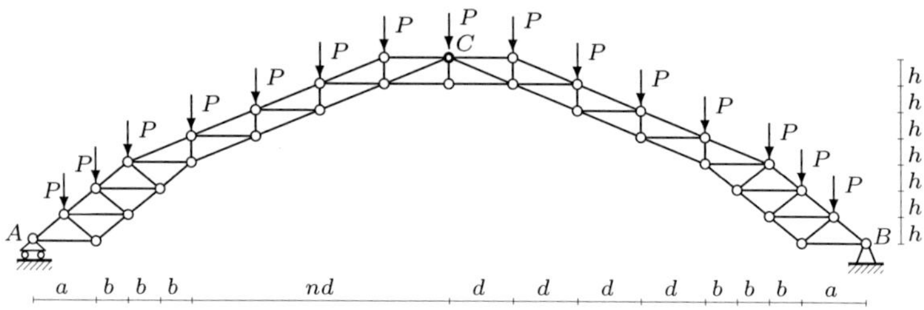

Fig. 23. Truss, $m = 3$, $n = 4$

Deflection. The coefficients in (1.15):

$$C_1 = (24n^3 + (72 + 36m)n^2 + 2(m+2)(13m+14)n+$$
$$+7m^3 + 8 + 32m + 26m^2)/32,$$

$$C_2 = m(5m^3 + 4(5+2n)m^2 + (24n+43)m + 40 + 40n)/384,$$

$$C_3 = (10n^4 + 4(4+5m)n^3 + 2(9m^2 + 12m + 1)n^2+$$
$$+2(3m^3 + 3m^2 - m + 4)n - 3m^3 - 6m^2 + 12)/24,$$

$$C_4 = (4n^3 + 6(m+2)n^2 + 4m(m+2)n + m^3 + 2m^2 - 4m - 12)/8.$$

Support offset. The coefficients in (1.16):

$$A_1 = (7m^3 + 4(4+5n)m^2 + 12(1+2n^2+4n)m + 4n(1+6n+3n^2))/8,$$

$$A_2 = m(5m^3 + 4(3+2n)m^2 + (12n+13)m + 4n + 6)/96,$$

$$A_3 = (3(2n-1)m^3 + 3(5n^2 + n - 2)m^2 + 2n(7n^2+$$
$$+6n - 4)m + n(n-1)(5n^2 + 9n + 1))/6,$$

$$A_4 = (m^3 + (1+3n)m^2 + 2(2n^2 + 2n - 3)m+$$
$$+2(n-1)(n^2 + 3n + 1))/2.$$

Forces:

$$O = -P(am^2 + 2a(n+1)m + 2n(n+2)a)/(4h),$$

$$D = -Pg/(2h),\ U = P(am^2 + 2a(n+1)m + 2(n+1)^2a)/(4h),$$

$$V = P(m^2 + 2(n+1)m + 2n^2 + 4n - 4)/4.$$

Supports reactions: $Y_A = Y_B = P(2n + 2m + 1)/2$, $X_B = 0$.

Truss 1.11

A truss with a height of $h(n + 1)$ (Fig. 24) containing $2n + 4$ panels, consists of $8n + 11$ rods.

1.11.1. Concentrated force in the mid-span

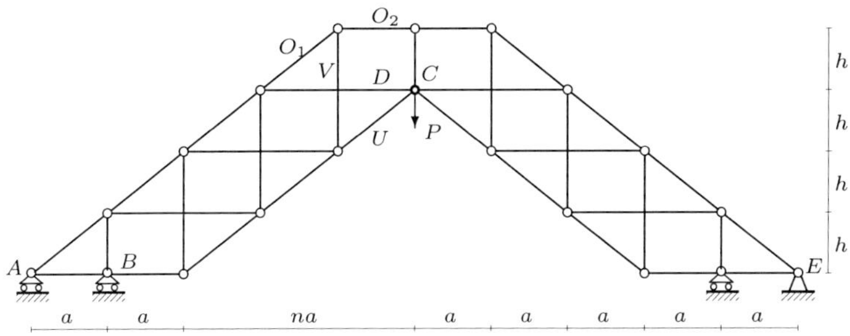

Fig. 24. Truss, $n = 3$

The truss is unchangeable when $n = 1, 3, 5, ...$ or when $n = 2k - 1$, $k = 1, 2, ...$. With $n = 2, 4, 6, ...$ the truss is kinematically variable.

$$\Delta = P(C_1 a^3 + C_2 c^3 + C_3 h^3)/(h^2 EF), \qquad (1.17)$$

where $c = \sqrt{a^2 + h^2}$.

The total length of the rods in the truss is

$$L_{sum} = 2(4k + 1)a + 2(4k - 1)c + (8k - 1)h.$$

Support A offset:

$$\delta_A = P(A_1 a^3 + A_2 c^3 + A_3 h^3)/(ahEF). \qquad (1.18)$$

Deflection. The coefficients in (1.17):

$C_1 = k(4k^2 + 5k + 2)/2$, $C_2 = k(4k^2 + 3k + 2)/6$,

$C_3 = (4k^3 + 3k^2 - 1)/2$.

Support offset. The coefficients in (1.18):

$A_1 = k(4k^2 + k - 1)$, $A_2 = k(2k - 1)(2k + 1)/3$,

$A_3 = k(4k^2 - k - 1)$.

Forces:

$O_1 = -Pck/(2h)$, $O_2 = -Pak/(2h)$,

$D = Pa(k+1)/(2h)$, $U = -Pc/(2h)$, $V = Pk/2$.

Supports reactions:

$Y_A = -Pk/2$, $Y_B = P(k+1)/2$, $X_E = 0$.

1.11.2. Load on the upper belt

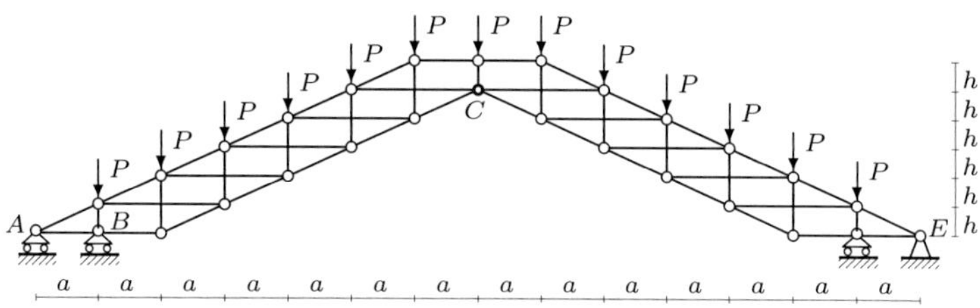

Fig. 25. Truss, $n = 5$

Deflection. The coefficients in (1.17):

$C_1 = k^2(4k+3)(2k+1)/2$, $C_2 = k^2(2k+1)(4k+1)/6$,

$C_3 = (8k^4 + 6k^3 + k^2 - 1)/2$.

Support offset. The coefficients in (1.18):

$A_1 = k(8k^3 + 2k^2 - 3k + 1)$, $A_2 = k^2(8k^2 - 5)/3$,

$A_3 = k(2k+1)(4k^2 - 3k + 1)$.

Forces:

$O_1 = -Pck(2k+1)/(2h)$, $O_2 = -Pak(2k+1)/(2h)$,

$D = Pa(2k^2 + k + 1)/(2h)$, $U = -Pc/(2h)$, $V = (2k^2 + k - 2)/2$.

Supports reactions:

$Y_A = -Pk(2k-1)/2$, $Y_B = P(k+1)(2k+1)/2$, $X_E = 0$.

Truss 1.12

A truss with a height of $h(n+1)$ (Fig. 26) containing $2n+6$ panels, consists of $8n+17$ rods.

The truss is unchangeable when $n = 2, 5, 8, \ldots$ or when $n = 3k-1$, $k = 1, 2, 3, \ldots$. With $n = 1, 3, 4, 6, 7, \ldots$ the truss is kinematically variable (Fig. 27).

1.12.1. Concentrated force in the mid-span

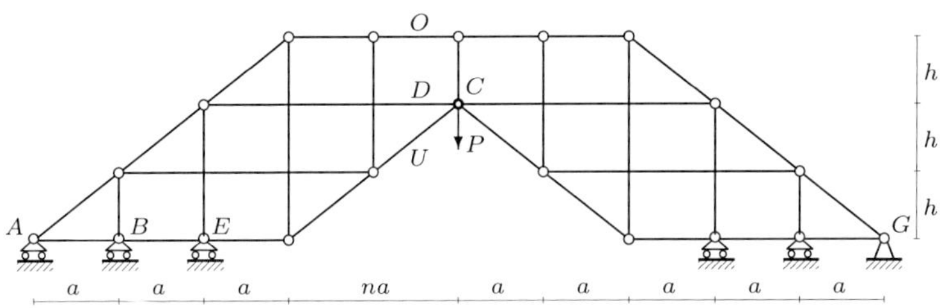

Fig. 26. Truss, $k = 1$, $n = 2$

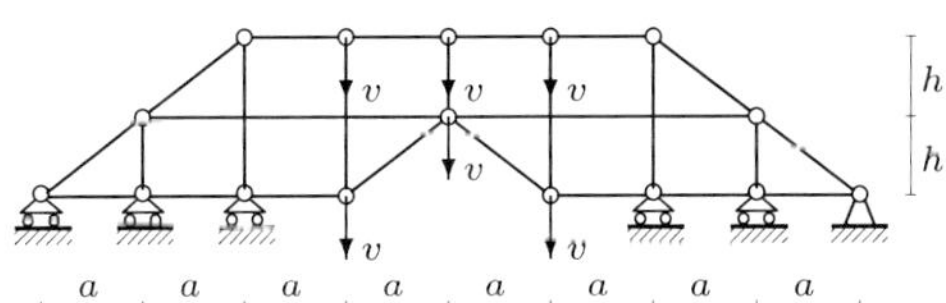

Fig. 27. Diagram of possible velocity of nodes for an instantly changeable truss, $n = 1$

The deflection (vertical displacement of the middle node C in the lower zone) has the form

$$\Delta = P(C_1a^3 + C_2c^3 + C_3h^3)/(h^2EF), \tag{1.19}$$

where $c = \sqrt{a^2+h^2}$.

The total length of the rods in the truss is

$$L_{sum} = 2(9k+2)a + 2(6k-1)c + (18k-1)h.$$

Support A offset:

$$\delta_A = P(A_1 a^3 + A_2 c^3 + A_3 h^3)/(ahEF). \tag{1.20}$$

Deflection. The coefficients in (1.19):

$C_1 = k(6k^2 + 8k + 3)/2$, $C_2 = k(2k^2 + 2k + 1)/2$,

$C_3 = (6k^3 + 5k^2 + k - 1)/2$.

Support offset. The coefficients in (1.20):

$A_1 = 2k(3k^2 + k - 1)$, $A_2 = k(4k^2 + k - 1)/2$,

$A_3 = k(3k + 1)(2k - 1)$.

Forces:

$O = -Pak/(2h)$, $D = Pa(k + 1)/(2h)$, $U = -Pc/(2h)$.

Supports reactions:

$Y_A = -Pk/2$, $Y_B = 0$, $Y_E = P(k + 1)/2$, $X_G = 0$.

1.12.2. Load on the upper belt

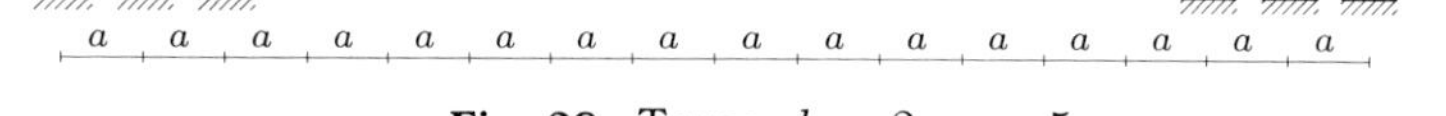

Fig. 28. Truss, $k = 2$, $n = 5$

Deflection. The coefficients in (1.19):

$C_1 = k^2(6k + 5)(3k + 2)/2$, $C_2 = k^2(11k^2 + 12k + 3)/4$,

$C_3 = (18k^4 + 18k^3 + 5k^2 + k - 1)/2$.

Support offset. The coefficients in (1.20):

$A_1 = k(18k^3 + 9k^2 - 8k - 1)$,

$A_2 = k(11k^3 + 4k^2 - 6k - 1)/2$, $A_3 = 2k(9k^3 - 2k + 1)$.

Forces:

$O = -Pak(3k+2)/(2h)$, $D = Pa(3k^2+2k+1)/(2h)$,

$U = -Pc/(2h)$.

Supports reactions:

$Y_A = -3Pk^2/2$, $Y_B = P(k+1)$, $Y_E = (k+1)(3k+1)/2$, $X_G = 0$.

Truss 1.13

A truss with a height of $h(n+1)$ (Fig. 29) containing $2n+8$ panels, consists of $8n+23$ rods.

1.13.1. Concentrated force in the mid-span

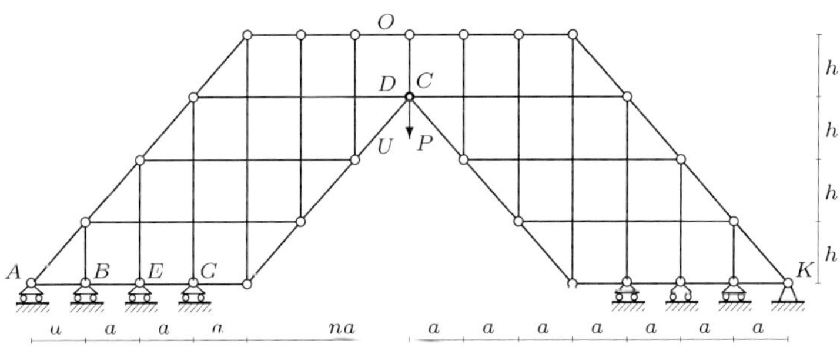

Fig. 29. Truss, $n = 3$

The truss is unchangeable when $n = 3, 7, 11, ...$ or when $n = 4k-1$, $k = 1, 2, 3, ...$. With $n = 1, 2, 4, 5, 6, 8, ...$ the truss is kinematically variable (Fig. 30).

The total length of the rods in the truss is

$$L_{sum} = 2(16k+3)a + 2(8k-1)c + (32k-1)h,$$

where $c = \sqrt{a^2+h^2}$.

The deflection (vertical displacement of the middle node C in the lower zone) has the form

$$\Delta = P(C_1a^3 + C_2c^3 + C_3h^3)/(h^2EF). \quad (1.21)$$

Support A offset:

$$\delta_A = P(A_1a^3 + A_2c^3 + A_3h^3)/(ahEF). \tag{1.22}$$

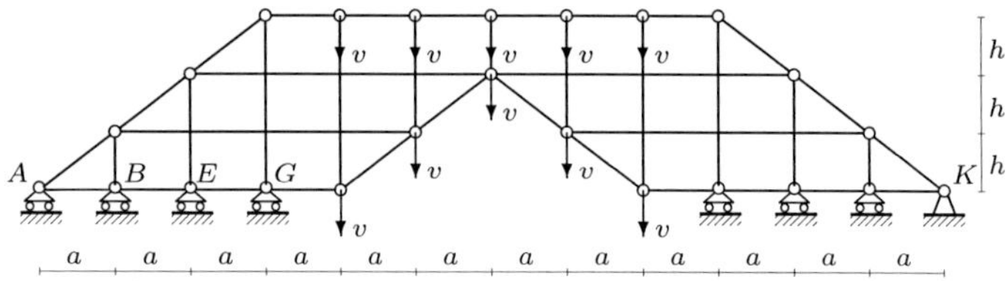

Fig. 30. Diagram of possible velocity of nodes for an instantly changeable truss, $n = 2$

Deflection. The coefficients in (1.21):

$C_1 = k(8k^2 + 11k + 4)/2$, $C_2 = k(8k^2 + 9k + 4)/6$,

$C_3 = (8k^3 + 7k^2 + 2k - 1)/2$.

Support offset. The coefficients in (1.22):

$A_1 = k(8k^2 + 3k - 3)$, $A_2 = k(8k^2 + 3k - 2)/3$,

$A_3 = k(8k^2 - k - 1)$.

Forces:

$O = -Pak/(2h)$, $D = Pa(k + 1)/(2h)$, $U = -Pc/(2h)$.

Supports reactions:

$Y_A = -Pk/2$, $Y_B = Y_E = 0$, $Y_G = P(k + 1)/2$, $X_K = 0$.

1.13.2. Load on the upper belt

Deflection. The coefficients in (1.21):

$C_1 = k^2(8k + 7)(4k + 3)/2$, $C_2 = k^2(14k + 11)(2k + 1)/6$,

$C_3 = (32k^4 + 36k^3 + 11k^2 + 2k - 1)/2$.

Support offset. The coefficients in (1.22):

$A_1 = k(32k^3 + 20k^2 - 15k - 5)$, $A_2 = k(28k^3 + 16k^2 - 13k - 4)/3$,

$A_3 = k(32k^3 + 4k^2 - 9k + 3)$.

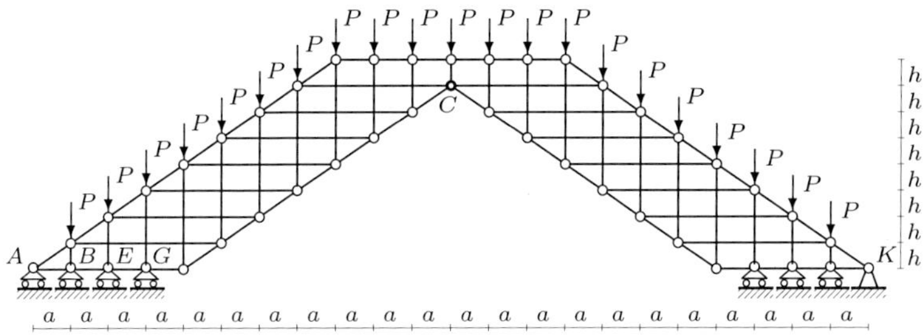

Fig. 31. Truss, $k = 2$, $n = 7$

Forces:

$O = -Pak(4k+3)/(2h)$, $D = Pa(4k^2+3k+1)/(2h)$,

$U = -Pc/(2h)$.

Supports reactions: $Y_A = -Pk(1+4k)/2$, $Y_B = Y_E = P(k+1)$,

$Y_G = P(k+1)(4k+1)/2$, $X_K = 0$.

Truss 1.14

A truss with a height of $(n+1)h$ consists of $8n+7$ rods (Fig. 32). The truss is unchangeable when $n = 2, 4, 6, 8, \ldots$ or when $n = 2k$, $k = 1, 2, \ldots$. With $n = 1, 3, 5, 7\ldots$ the truss is kinematically variable (Fig. 33).

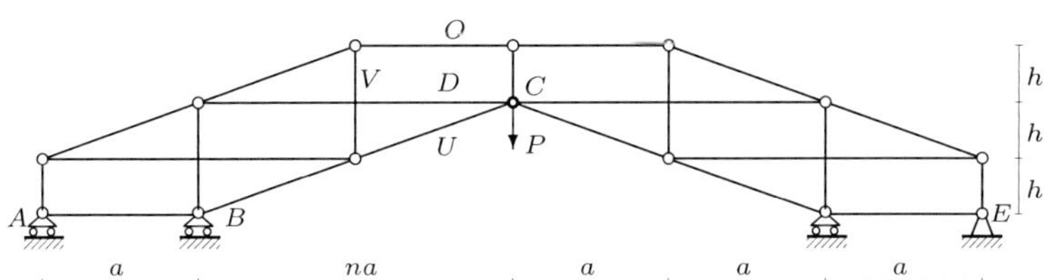

Fig. 32. Truss, $n = 2$

The deflection (vertical displacement of the middle node C in the lower zone) has the form

$$\Delta = P(C_1a^3 + C_2c^3 + C_3h^3)/(h^2EF), \qquad (1.23)$$

where $c = \sqrt{a^2 + h^2}$.

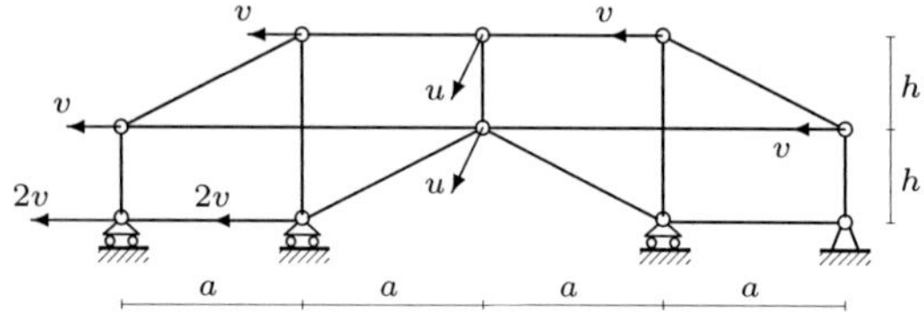

Fig. 33. Diagram of possible velocity of nodes for an instantly changeable truss, $n = 1$, $v/h = u/c$, $c = \sqrt{a^2 + h^2}$

The total length of the rods in the truss is

$$L_{sum} = 4(2k + 1)a + 8ck + (8k + 3)h.$$

Support A offset:

$$\delta_A = P(A_1 a^3 + A_2 c^3 + A_3 h^3)/(ahEF). \tag{1.24}$$

1.14.1. Concentrated force in the mid-span

Deflection. The coefficients in (1.23):

$C_1 = C_3 = k(4k^2 + 5k + 2)/2$,

$C_2 = k(4k^2 + 3k + 2)/6$.

Support offset. The coefficients in (1.24):

$A_1 = A_3 = k(4k^2 + 5k + 2)$, $A_2 = k(k + 1)(4k - 1)/3$.

Forces:

$O = -Pak/(2h)$, $D = Pa(k + 1)/(2h)$,

$U = -Pc/(2h)$, $V = Pk/2$.

Supports reactions:

$Y_A = Y_E = -Pk/2$,

$Y_B = P(k + 1)/2$, $X_E = 0$.

1.14.2. Load on the upper belt

Deflection. The coefficients in (1.23):

$C_1 = k^2(4k + 3)(2k + 1)/2$, $C_2 = k^2(2k + 1)(4k + 1)/6$,

$C_3 = k(8k^3 + 10k^2 + 5k + 2)/2.$

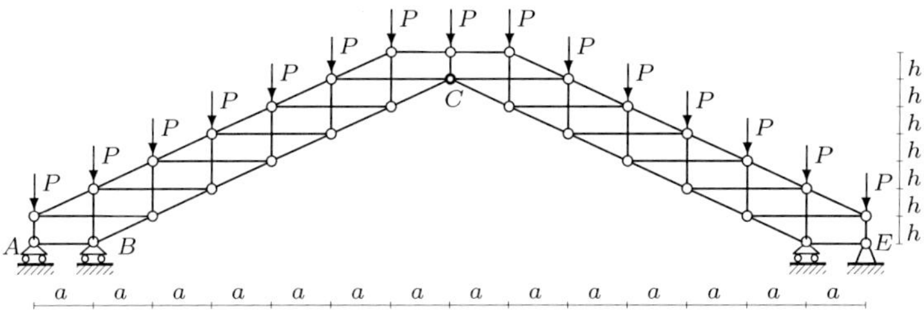

Fig. 34. Truss, $n = 6$

Support offset. The coefficients in (1.24):

$A_1 = k^2(4k+3)(2k+1)$, $A_2 = 2k(4k^2-1)(k+1)/3$,

$A_3 = k(8k^3 + 10k^2 + 5k + 2)$.

Forces:

$O = -Pak(1+2k)/(2h)$, $D = Pa(2k^2+k+1)/(2h)$,

$U = -Pc/(2h)$, $V = P(2k^2+k-2)/2$.

Supports reactions:

$Y_A = Y_E = P(2+k-2k^2)/2$, $Y_B = P(k+1)(1+2k)/2$, $X_E = 0$.

Truss 1.15

A truss with a height of $(n+1)h$ consists of $8n+13$ rods (Fig. 35).

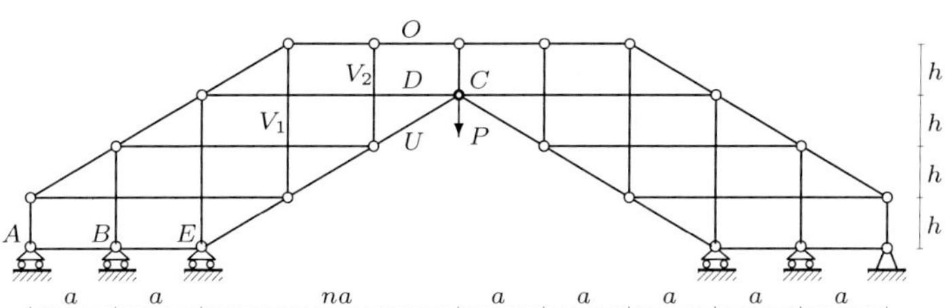

Fig. 35. Truss, $n = 3$

The truss is unchangeable when $n = 3, 6, 9, \ldots$ or when $n = 3k$, $k = 1, 2, \ldots$. With $n = 1, 2, 4, 5\ldots$ the truss is kinematically variable (Fig. 36).

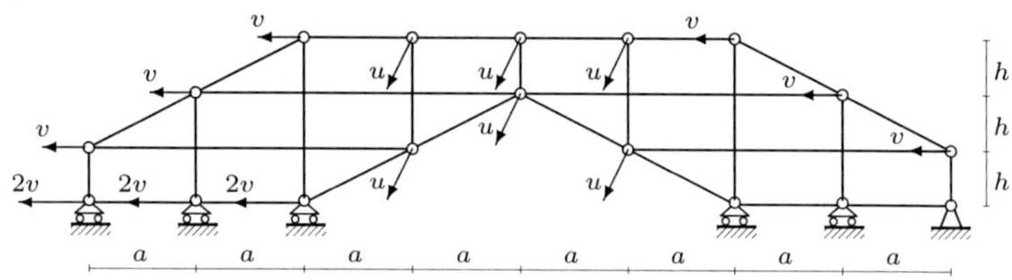

Fig. 36. Diagram of possible velocity of nodes for an instantly changeable truss, $n = 2$, $v/h = u/c$, $c = \sqrt{a^2 + h^2}$

The deflection (vertical displacement of the middle node C in the lower zone) has the form

$$\Delta = P(C_1 a^3 + C_2 c^3 + C_3 h^3)/(h^2 EF), \tag{1.25}$$

where $c = \sqrt{a^2 + h^2}$.

The total length of the rods in the truss is

$$L_{sum} = 2(4 + 9k)a + 12ck + (5 + 18k)h.$$

Support A offset:

$$\delta_A = P(A_1 a^3 + A_2 c^3 + A_3 h^3)/(ahEF). \tag{1.26}$$

1.15.1. Concentrated force in the mid-span

Deflection. The coefficients in (1.25):

$C_1 = k(6k^2 + 8k + 3)/2$, $C_2 = k(2k^2 + 2k + 1)/2$,

$C_3 = k(6k^2 + 7k + 3)/2$.

Support offset. The coefficients in (1.26):

$A_1 = k(6k^2 + 8k + 3)$, $A_2 = k(k + 1)(4k - 1)/2$,

$A_3 = k(6k^2 + 7k + 3)$.

Forces:

$O = -Pak/(2h)$, $D = Pa(k + 1)/(2h)$, $U = -Pc/(2h)$,

$V_1 = Pk/2$, $V_2 = 0$.

Supports reactions:

$Y_A = -Pk/2$, $Y_B = 0$, $Y_E = P(k+1)/2$.

1.15.2. Load on the upper belt

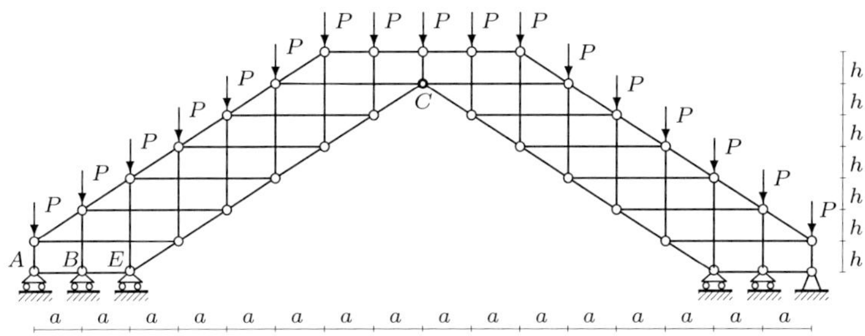

Fig. 37. Truss, $n = 6$

Deflection. The coefficients in (1.25):

$C_1 = k^2(6k+5)(3k+2)/2$, $C_2 = k^2(11k^2+12k+3)/4$,

$C_3 = k(9k^3+12k^2+6k+2)$.

Support offset. The coefficients in (1.26):

$A_1 = k^2(6k+5)(3k+2)$, $A_2 = k(k+1)(11k^2+k-3)/2$,

$A_3 = 2k(9k^3+12k^2+6k+2)$.

Forces:

$O = -Pak(2+3k)/(2h)$, $D = Pa(3k^2+2k+1)/(2h)$,

$U = -Pc/(2h)$, $V_1 = P(3k^2+2k-2)/2$, $V_2 = -P$.

Supports reactions:

$Y_A = P(2-3k^2)/2$, $Y_B = P(k+1)$, $Y_E = P(k+1)(3k+1)/2$.

Truss 1.16

A truss with a height of $h(2m+1)$ consists of $8(m+n)+11$ rods (Fig. 38), $b = a/2$. The deflection (vertical displacement of the middle node C in the upper zone) has the form

$$\Delta = P(C_1a^3 + C_2c^3 + C_3d^3 + C_4f^3 + C_5g^3)/(h^2EF), \qquad (1.27)$$

where $c = \sqrt{a^2+h^2}$, $d = \sqrt{a^2+16h^2}$, $f = \sqrt{25a^2+16h^2}$, $g = \sqrt{9a^2+16h^2}$.

The total length of the rods in the truss is

$$L_{sum} = 2(4n+5)a + 2(2n+3)c + (2m-1)d + (m-1/2)g + fm.$$

Support A offset:

$$\delta_A = P(A_1a^3 + A_2c^3 + A_3d^3 + A_4f^3 + A_5g^3)/(ahEF). \qquad (1.28)$$

1.16.1. Concentrated force in the mid-span

Deflection. The coefficients in (1.27):

$$C_1 = (3(105+128n)m^2 + 12(245+128n^2+3(-1)^m+384n)m+$$
$$+8(891+21(-1)^m+1760n+1152n^2+256n^3))/768,$$

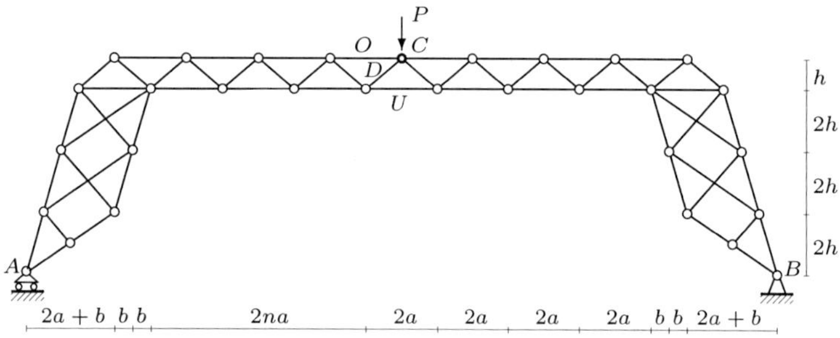

Fig. 38. Truss, $m = n = 3$

$$C_2 = (16n+24+m^2+8m)/16,$$

$$C_3 = (2m^3+12m^2+49m+6-6(-1)^m)/3072,$$

$$C_4 = (2m+1-(-1)^m)/4096,$$

$$C_5 = (2m-1+(-1)^m)/4096.$$

Support offset. The coefficients in (1.28):

$$A_1 = ((105+128n)m^2 + 4(64n^2+200n+3(-1)^m+128)m+$$

$$+32(4+(-1)^m+6n+2n^2))/32,$$

$$A_2 = m(m+4)/2,\ A_3 = (2m^3+6m^2+m+3-3(-1)^m)/384,$$

$$A_4 = (2m+1-(-1)^m)/512,$$

$$A_5 = (2m-1+(-1)^m)/512.$$

Forces:

$$O = -Pa(4n+m+4)/(4h),\ D = -Pc/(2h),$$

$$U = Pa(4n+m+6)/(4h).$$

Supports reactions: $Y_A = Y_B = P/2$, $X_B = 0$.

1.16.2. Load on the upper belt

Deflection. The coefficients in (1.27):

$$C_1 = (3(3+2n)(128n+105)m^2+4(3+2n)(320n^2+$$
$$+960n+9(-1)^m+607)m+8(320n^4+1920n^3+$$
$$+4384n^2+6(745+7(-1)^m)n+1713+63(-1)^m))/768,$$

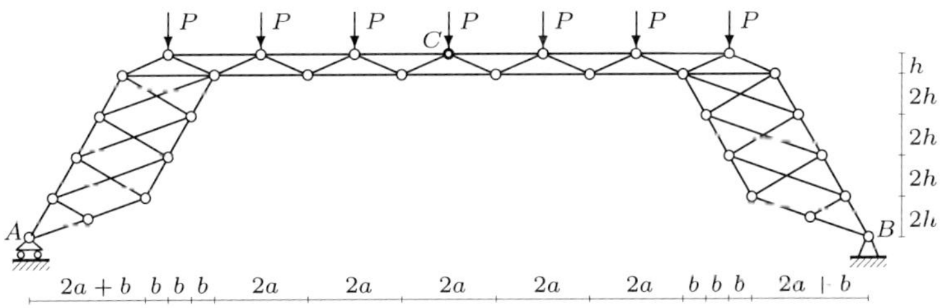

Fig. 39. Truss, $m = 4$, $n = 2$

$$C_2 = ((2n+3)m^2+4(4n+5)m+16n^2+48n+40)/16,$$

$$C_3 = (2n+3)(12m^2+2m^3-6(-1)^m+6+49m)/3072,$$

$$C_4 = (2n+3)(2m+1-(-1)^m)/4096,$$

$$C_5 = (2n+3)(2m-1+(-1)^m)/4096.$$

Support offset. The coefficients in (1.28):

$$A_1 = (3+2n)(128(1+4m)n^2+96(m+4)(1+4m)n+$$

$$+96(-1)^m + 1024m + 36(-1)^m m + 315m^2 + 256)/96,$$

$$A_2 = m((2m + 8)n + 3m + 8)/2,$$

$$A_3 = (3 + 2n)(6m^2 + m + 2m^3 - 3(-1)^m + 3)/384,$$

$$A_4 = (3 + 2n)(2m + 1 - (-1)^m)/512,$$

$$A_5 = (3 + 2n)(2m - 1 + (-1)^m)/512.$$

Forces:

$$O = -Pa(4n^2 + 2(m + 6)n + 3m + 8)/(4h),$$

$$D = -Pc/(2h),\ U = Pa(4n^2 + 2(6 + m)n + 10 + 3m)/(4h).$$

Supports reactions:

$$Y_A = Y_B = (2n + 3)P/2,\ X_B = 0.$$

Truss 1.17

A truss with a height of $(m + 1)h$ (Fig. 40) containing $2n + 2m$ panels, consists of $8(n + m) + 7$ rods [45]. The total length of the rods in the truss is

$$L_{sum} = 2(3 + 2m + 4n)a + 2(2 + 3m + 2n)c,$$

where $c = \sqrt{a^2 + h^2}$.

The deflection (vertical displacement of the middle node C in the lower zone) has the form

$$\Delta = P(C_1 a^3 + C_2 c^3)/(h^2 EF). \quad (1.29)$$

Support A offset:

$$\delta_A = P(A_1 a^3 + A_2 c^3)/(ahEF). \quad (1.30)$$

1.17.1. Concentrated force in the mid-span

Deflection. The coefficients in (1.29):

$$C_1 = (16n^3 + 24(m + 2)n^2 + 2(6m^2 + 24m + 25)n +$$
$$+3m^2 + 18m + 18)/6,$$

$$C_2 = (m^3 + 6m^2 + 17m + 12n + 12)/12.$$

Case $m = n$:

$C_1 = (52n^3 + 99n^2 + 68n + 18)/6,$

$C_2 = (n^3 + 6n^2 + 29n + 12)/12.$

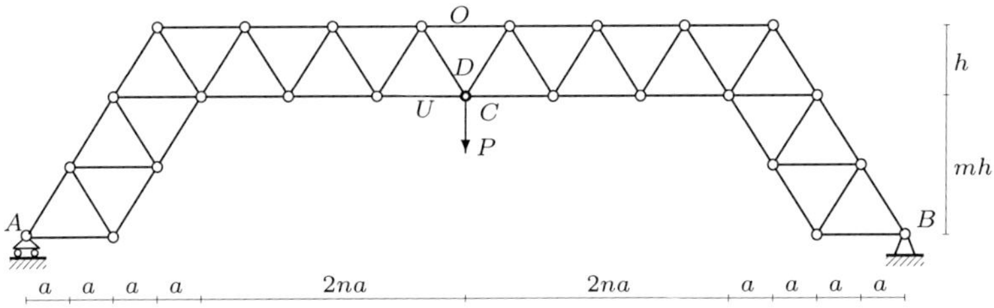

Fig. 40. Truss, $m = 2$, $n = 3$

Support offset. The coefficients in (1.30):

$A_1 = (4n + 1)m^2 + 2(n + 2)(2n + 1)m + 2(n + 1)^2,$

$A_2 = m(2m^2 + 9m + 13)/12.$

Case $m = n$:

$A_1 = 8n^3 + 13n^2 + 8n + 2,$

$A_2 = n(2n^2 + 9n + 13)/12.$

Forces:

$O = -Pa(m + 2 + 2n)/(2h)$, $D = Pc/(2h)$,

$U = Pa(2n + m + 1)/(2h)$.

Supports reactions:

$Y_A = Y_B = P/2$, $X_B = 0$.

1.17.2. Load on the upper belt

Deflection. The coefficients in (1.29):

$$C_1 = (20n^4 + 40(m + 2)n^3 + 4(9m^2 + 33m + 31)n^2 + \\ +2(6m^3 + 33m^2 + 67m + 44)n + 3(m + 4)(m + 2)(m + 1))/6,$$

$$C_2 = (48n^2 + 8(m + 1)(m^2 + 5m + 12)n + 5m^4 + 36m^3 +$$

$+115m^2 + 156m + 48)/48.$

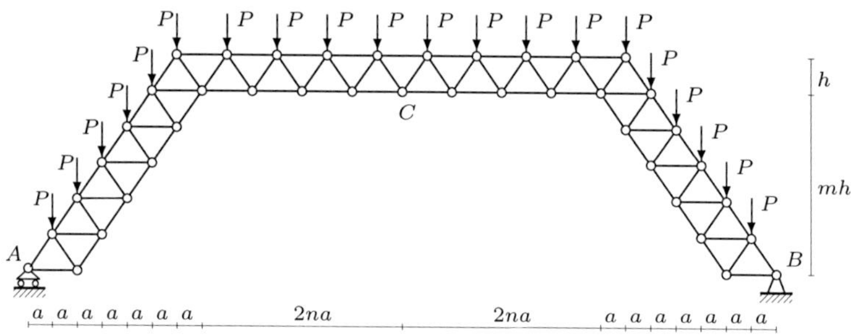

Fig. 41. Truss, $n = 4$, $m = 5$

Case $m = n$:

$C_1 = (n + 1)(108n^3 + 173n^2 + 106n + 24)/6,$

$C_2 = (13n^4 + 84n^3 + 299n^2 + 252n + 48)/48.$

Support offset. The coefficients in (1.30):

$A_1 = (1 + 4n)m^3 + (8n^2 + 16n + 5)m^2+$

$+4(4n + 3)(n + 2)(n + 1)m/3 + 4(n + 1)(2n^2 + 4n + 3)/3,$

$A_2 = m(5m^3 + 4(2n + 7)m^2 + (36n + 55)m + 4(13n + 8))/24.$

Case $m = n$:

$A_1 = (n + 1)(52n^3 + 67n^2 + 40n + 12)/3,$

$A_2 = n(13n^3 + 64n^2 + 107n + 32)/24.$

Forces:

$O = -U = -Pa(2n^2 + 2(m + 2)n + (m + 1)(m + 2))/(2h)$, $D = 0$.

Supports reactions:

$Y_A = Y_B = (n + m + 1)P$, $X_B = 0$.

Chapter 2

Frames

Truss 2.1

A truss with a height of $(m+1)h$ consists of $8n+8m+17$ rods (Fig. 42). The total length of the rods in the truss is

$$L_{sum} = (5+4n)a + (4m+2n+7)c + 2(2m+1)h,$$

where $c = \sqrt{a^2+h^2}$.

The deflection (vertical displacement of the middle node C in the upper zone) has the form

$$\Delta = P(C_1a^3 + C_2c^3 + C_3h^3)/(h^2EF). \tag{2.1}$$

Support A offset:

$$\delta_A = P(A_1a^3 + A_2c^3 + A_3h^3)/(ahEF). \tag{2.2}$$

2.1.1. Concentrated force in the mid-span

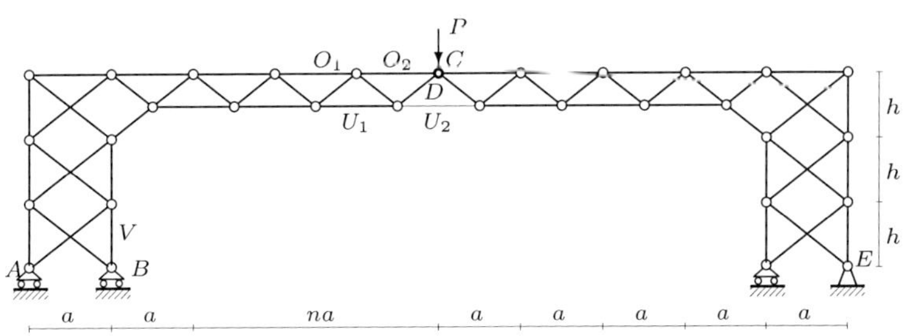

Fig. 42. Truss, $m=2$, $n=3$

Deflection. The coefficients in (2.1):

$C_1 = (8n^3+n+6)/6$, $C_2 = (n+2)/2$, $C_3 = (5m+1)/2$.

Support offset. The coefficients in (2.2):

$A_1 = 3 + (-1)^m - (3 + 2(-1)^m)n^2$, $A_2 = ((-1)^m + 5)/4$,

$A_3 = (6(-1)^m m - (-1)^m + 6m^2 + 18m + 5)/4$.

Forces:

$O_1 = -Pa(2n-3)/(2h)$, $O_2 = -Pa(2n-1)/(2h)$,

$D = -Pc/(2h)$, $U_1 = Pa(n-1)/h$, $U_2 = Pan/h$, $V = -P$.

Supports reactions:

$Y_A = Y_E = -P/2$, $Y_B = P$, $X_E = 0$.

2.1.2. Load on the upper belt

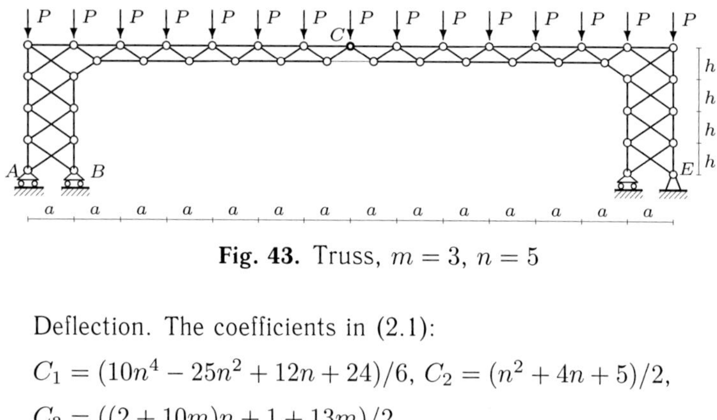

Fig. 43. Truss, $m = 3$, $n = 5$

Deflection. The coefficients in (2.1):

$C_1 = (10n^4 - 25n^2 + 12n + 24)/6$, $C_2 = (n^2 + 4n + 5)/2$,

$C_3 = ((2 + 10m)n + 1 + 13m)/2$.

Support offset. The coefficients in (2.2):

$A_1 = (-4(2(-1)^m + 3)n^3 + (57 + 32(-1)^m)n + 21(-1)^m + 45)/3$,

$A_2 = (2((-1)^m + 5)n + 5(-1)^m + 17)/4$,

$A_3 = (2(7 + 6n)m^2 + 2(7 + 6n)((-1)^m + 3)m-$

$-5(-1)^m - 2(-1)^m n + 9 + 10n)/4$.

Forces:

$O_1 = -Pa(2n^2 - 9)/(2h)$, $O_2 = -Pa(2n^2 - 5)/(2h)$,

$D=-Pc/(2h)$,

$U_1=Pa(n^2-3)/h$, $U_2=Pa(n^2-2)/h$, $V=-P(2n+3)$.

Supports reactions:

$Y_A=Y_E=-P(2n+1)/2$, $Y_B=P(2n+3)$, $X_A=0$.

Truss 2.2

A truss with a height of $2mh$ (Fig. 44) containing $2(n+m)$ panels, consists of $16m+8n-7$ rods. The total length of the rods in the truss is

$$L_{sum}=2(2m+2n-1)a+2(3m+n-1)c+(6m+2n-3)h,$$

where $c=\sqrt{a^2+h^2}$.

The deflection (vertical displacement of the middle node C in the lower zone) has the form

$$\Delta=P(C_1a^3+C_2c^3+C_3h^3)/(2h^2EF). \qquad (2.3)$$

Support A offset:

$$\delta_A=P(A_1a^3+A_2c^3+A_3h^3)/(ahEF). \qquad (2.4)$$

2.2.1. Concentrated force in the mid-span

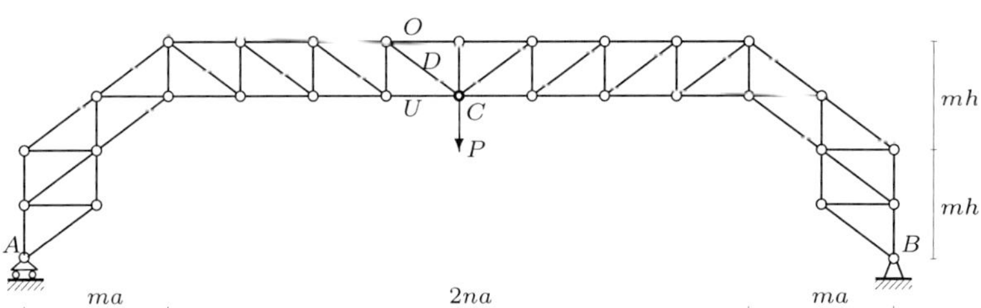

Fig. 44. Truss, $m=2$, $n=4$

Deflection. The coefficients in (2.3):

$C_1=(2n^3+6n^2m+(6m^2+1)n+3m)/3$,

$C_2=(3n+2m^3+m)/3$, $C_3=(3n+3m^2-3)/3$.

Support offset. The coefficients in (2.4):

$A_1 = (2(2m-1)n^2 + (8m^2-4m-1)n + 4m-3)/2,$

$A_2 = m(10m^2-6m-1)/6,$

$A_3 = (5m^2-5m+1)/2.$

Forces:

$O = -Pa(n+m)/(2h)$, $D = Pc/(2h)$, $U = Pa(n+m-1)/(2h)$.

Supports reactions:

$Y_A = Y_B = P/2$, $X_B = 0$.

2.2.2. Load on the upper belt

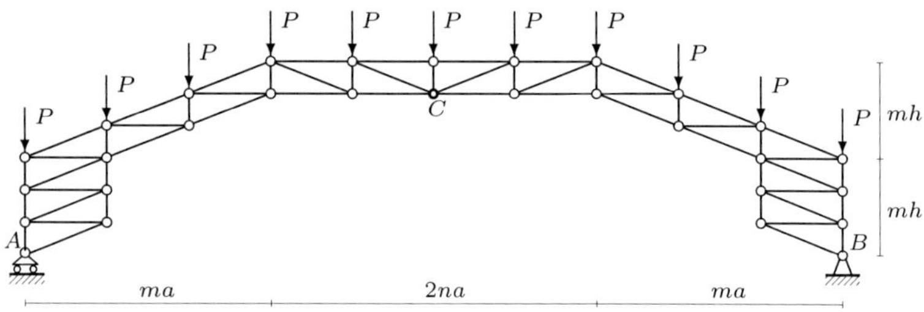

Fig. 45. Truss, $m = 3$, $n = 2$

Deflection. The coefficients in (2.3):

$C_1 = (5n^4 + 20n^3m + (30m^2+1)n^2 + 2m(5+6m^2)n + 6m^2)/6,$

$C_2 = (6n^2 + 4m(1+2m^2)n + m^2(1+5m^2))/6,$

$C_3 = n^2 + 2nm^2 + (m+1)(m^2+m-1).$

Support offset. The coefficients in (2.4):

$A_1 = (8(2m-1)n^3 + 3(16m^2-8m-1)n^2 +$

$+4(6m^3-3m^2+5m-4)n + 18m^2 - 24m + 9)/6,$

$A_2 = (2m(10m^2-6m-1)n + m(13m^3-8m^2-4m+2))/6,$

$A_3 = (2(5m^2-5m+1)n + 6m^3 - 4m + 1)/2.$

Forces:

$O = -Pa(n+m)^2/(2h)$, $D = Pc/(2h)$,

$U = Pa((n+m)^2 - 1)/(2h)$.

Supports reactions: $Y_A = Y_B = P(2n+2m+1)/2$, $X_B = 0$.

Truss 2.3

A truss with a height of $(2m+1)f + (k+1)h$ consists of $8(m+n+k)+11$ rods (Fig. 46), $d = a/2$. The total length of the rods in the truss is

$$L_{sum} = c(3k+2n+2) + a(2k+4n+3) + 2(4m+1)f + 2(2m+1)g,$$

where $c = \sqrt{a^2+4h^2}$, $g = \sqrt{a^2+f^2}$.

The deflection (vertical displacement of the middle node C in the lower zone) has the form

$$\Delta = P(C_1a^3 + C_2c^3 + C_3fh^2)/(h^2EF). \quad (2.5)$$

Support A offset, case $f = h$:

$$\delta_A = P(A_1a^3 + A_2c^3 + A_3h^3)/(ahEF). \quad (2.6)$$

2.3.1. Concentrated force in the mid-span

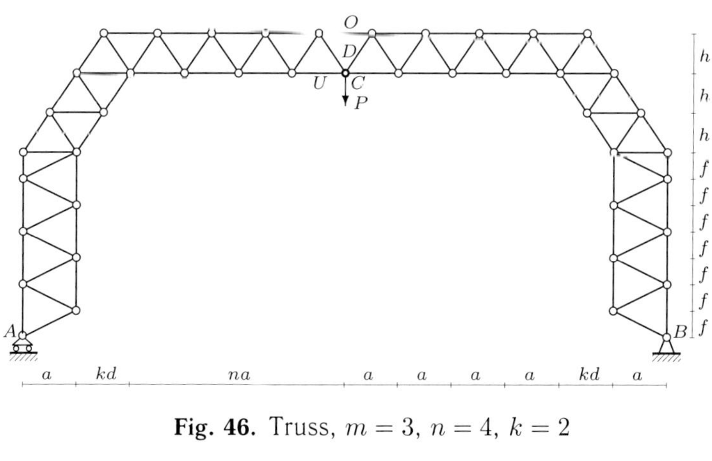

Fig. 46. Truss, $m = 3$, $n = 4$, $k = 2$

Deflection. The coefficients in (2.5):

$C_1 = (16n^3+24(k+2)n^2+2(25+6k^2+24k)n+18k+18+3k^2)/48,$

$C_2 = (12n + (k+1)(k^2+5k+12))/96,$

$C_3 = (2m+1)/2.$

Support offset. The coefficients in (2.6), case $f = h$:

$A_1 = (2(3+2k+4m)n^2 + 2(k+2)(2k+3+4m)n+$
$+k^2+5k+3+6m+2mk)/4,$

$A_2 = (2k^3 + 6(m+2)k^2 + (25+24m)k + 24m + 12)/48,$

$A_3 = 2m^2 + 2m + 1.$

Forces:

$O = -Pa(2n+k+2)/(4h)$, $U = Pa(2n+k+1)/(4h)$,

$D = Pc/(4h)$.

Supports reactions: $Y_A = Y_B = P/2$, $X_B = 0$.

2.3.2. Load on the upper belt

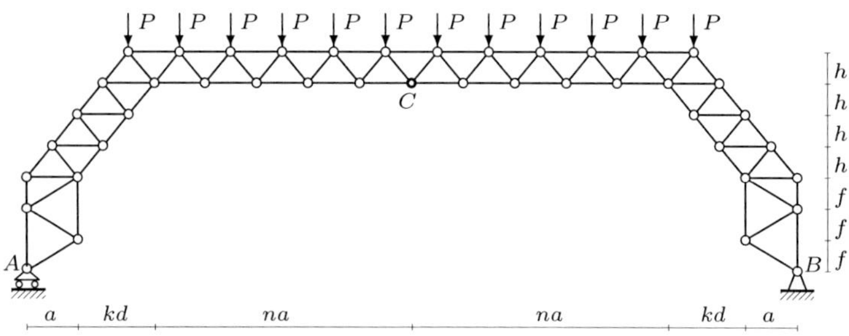

Fig. 47. Truss, $m = 1$, $n = 5$, $k = 3$

Deflection. The coefficients in (2.5):

$C_1 = (10n^4 + 20(2+k)n^3 + 2(31+30k+6k^2)n^2+$
$+(55k+44+15k^2)n + 3(k+4)(k+1))/24,$

$C_2 = (6n^2 + (k+1)(k^2+5k+12)n + k^3 + 6k^2 + 14k + 6)/48,$

$C_3 = (2m+1)(n+1).$

Support offset. The coefficients in (2.6), case $f = h$:

$$A_1 = (n+1)(3(4n+1)k^2 + 2(4n^2 + 12nm + 17n + 3m + 6)k + $$
$$+2(8n^2m + 16nm + 6n^2 + 12n + 6m + 3))/6,$$
$$A_2 = ((2n+2)k^3 + 6(n+1)(m+2)k^2 + (25n + 24nm + 24m + 19)k +$$
$$+6(2n+1)(2m+1))/24,$$
$$A_3 = 2(2m^2 + 2m + 1)(n+1).$$

Forces:

$O = -Pa(n+1)(n+k+1)/(2h)$, $D = 0$, $U = -O$.

Supports reactions: $Y_A = Y_B = (n+1)P$, $X_B = 0$.

Truss 2.4

A truss with a height of $(m+1)h$ (Fig. 48) containing $2(n+m+1)$ panels, consists of $8m + 8n + 5$ rods [49].

The deflection (vertical displacement of the middle node C in the lower zone) has the form

$$\Delta = P(C_1a^3 + C_2c^3 + C_3h^3)/(h^2EF), \qquad (2.7)$$

where $c = \sqrt{a^2 + h^2}$.

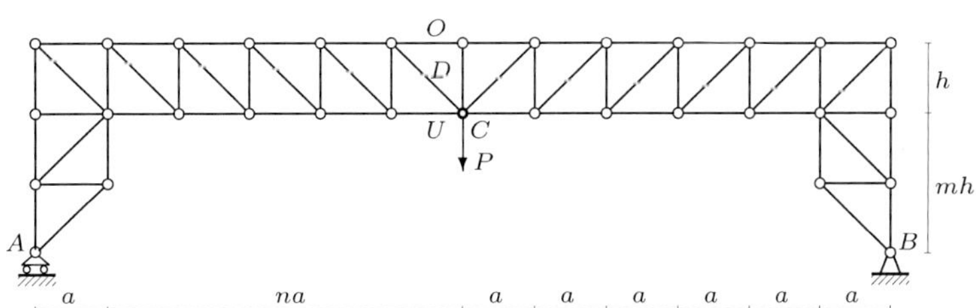

Fig. 48. Truss, $m = 2$, $n = 5$

The total length of the rods in the truss is

$$L_{sum} = 2(m + 2n + 1)a + 2(m + n + 1)c + (4m + 2n + 1)h.$$

Support A offset:

$$\delta_A = P(A_1 a^3 + A_2 c^3 + A_3 h^3)/(ahEF). \qquad (2.8)$$

2.4.1. Concentrated force in the mid-span

Deflection. The coefficients in (2.7):

$C_1 = (2n^3 + 6n^2 + 7n + 3)/6,$

$C_2 = (n + 1)/2,\ C_3 = (2n + 2m + 1)/4.$

Support offset. The coefficients in (2.8):

$A_1 = (2n^2m + n^2 + 4nm + n + 2m)/2,\ A_2 = m,\ A_3 = (m^2 + 3m - 1)/2.$

Forces:

$O = -Pa(n + 1)/(2h),\ D = Pc/(2h),\ U = Pan/(2h).$

Supports reactions:

$Y_A = Y_B = P/2,\ X_B = 0.$

2.4.2. Load on the upper belt

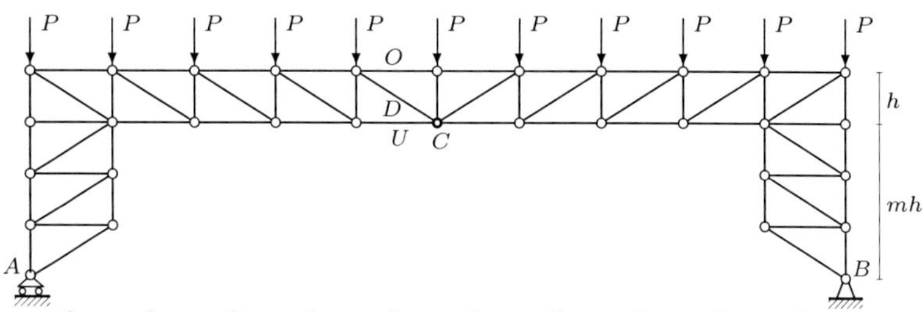

Fig. 49. Truss, $m = 3$, $n = 4$

Deflection. The coefficients in (2.7):

$C_1 = (5n^4 + 20n^3 + 31n^2 + 22n + 6)/12,\ C_2 = (1 + n^2 + 2n)/2,$

$C_3 = (2n^2 + 6(n + m) + 4nm + 3)/4.$

Support offset. The coefficients in (2.8):

$A_1 = (n + 1)(4n^2 + 8n^2m + 16nm + 5n + 6m)/6,$

$A_2 = m(2n+1)$, $A_3 = (m^2+3m-1)(2n+3)/2$.

Forces:

$O = -Pa(n+1)^2/(2h)$, $D = Pc/(2h)$,

$U = Pan(n+2)/(2h)$.

Supports reactions:

$Y_A = Y_B = P(2n+3)/2$, $X_B = 0$.

Truss 2.5

A truss with a height of $2(m+1)h$ (Fig. 50) containing $2(n+m)$ panels, consists of $8m+8n+13$ rods.

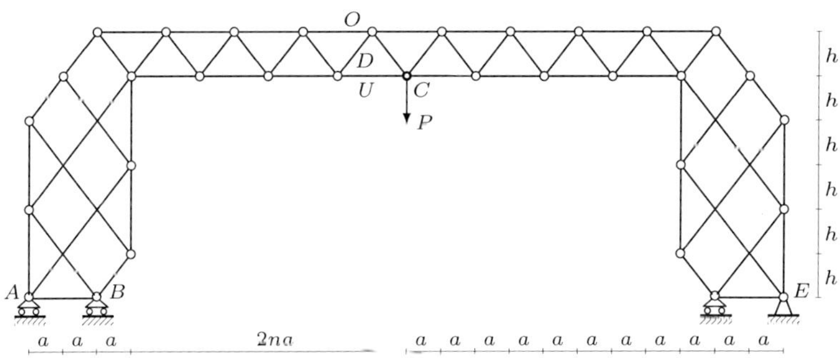

Fig. 50. Truss, $m = 2$, $n = 4$

The truss is unchangeable when $n = 2, 5, 8, 11, ...$ or when $m = 3k-1$, $k = 1, 2, ...$. With $m = 1, 3, 4, 6, 7...$ the truss is kinematically variable (Fig. 51).

The deflection (vertical displacement of the middle node C in the lower zone) has the form

$$\Delta = P(C_1a^3 + C_2c^3 + C_3h^3)/(h^2EF), \tag{2.9}$$

where $c = \sqrt{a^2+h^2}$.

The total length of the rods in the truss is

$$L_{sum} = 4(2n+1)a + 2(18k+2n-1)c + 8(3k-1)h.$$

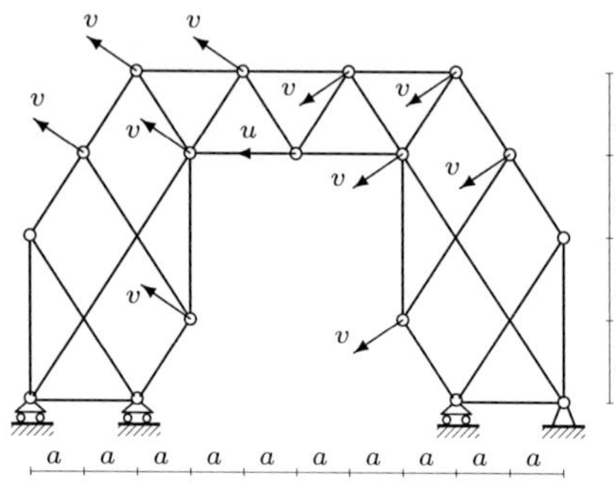

Fig. 51. Diagram of possible velocity of nodes for an instantly changeable truss, $m = n = 1$, $u/h = 2v/c$, $c = \sqrt{a^2 + h^2}$

Support A offset:

$$\delta_A = P(A_1 a^3 + A_2 c^3 + A_3 h^3)/(ahEF). \tag{2.10}$$

2.5.1. Concentrated force in the mid-span

Deflection. The coefficients in (2.9):

$C_1 = (16n^3 + 24n^2 + 14n + 3)/6,$

$C_2 = (n + 3k/2)$, $C_3 = 3k$.

Support offset. The coefficients in (2.10):

$A_1 = 2(4n^2 + 4n + 1)k - 2n^2 - 2n - 1,$

$A_2 = (12k^2 - 1)/2$, $A_3 = 6k^2$.

Forces:

$O = -Pa(2n + 1)/(2h)$, $D = Pc/(2h)$, $U = Pan/h$.

Supports reactions: $Y_A = 0$, $Y_B = P/2$, $X_E = 0$.

2.5.2. Load on the upper belt

Deflection. The coefficients in (2.9):

$C_1 = 2n(5n^3 + 10n^2 + 7n + 2)/3,$

$C_2 = n^2 + (m + 1)n + (2m + 1)/2$, $C_3 = 2(m + 1)(n + 1)$.

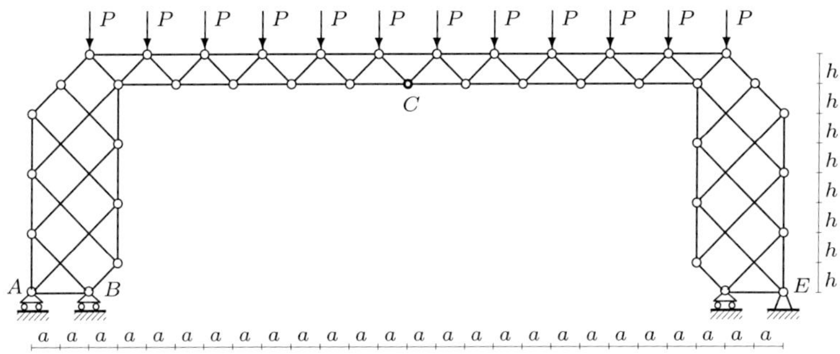

Fig. 52. Truss, $m = 3$, $n = 5$

Support offset. The coefficients in (2.10):

$A_1 = 4(4n(2n^2 + 3n + 1)k - 2n^3 - 3n^2 - n)/3,$

$A_2 = 12(n+1)k^2 - 2k - n,\ A_3 = 12(n+1)k^2.$

Forces:

$O = -Pan(n+1)/h,\ D = 0,\ U = Pan(n+1)/h.$

Supports reactions:

$Y_A = 0,\ Y_B = P(n+1),\ X_E = 0.$

Truss 2.6

A truss with a height of $(2m+1)h$ (Fig. 53) containing $2(n+m)$ panels, consists of $8m+8n+5$ rods. The total length of the rods in the truss is

$$L_{sum} = 2(4n+1)a + 2(4m+2n+1)c + 8mh,$$

where $c = \sqrt{a^2 + h^2}$.

The deflection (vertical displacement of the middle node C in the lower zone) has the form

$$\Delta = P(C_1 a^3 + C_2 c^3 + C_3 h^3)/(h^2 EF). \qquad (2.11)$$

Support A offset:

$$\delta_A = P(A_1 a^3 + A_2 c^3 + A_3 h^3)/(ahEF). \qquad (2.12)$$

2.6.1. Concentrated force in the mid-span

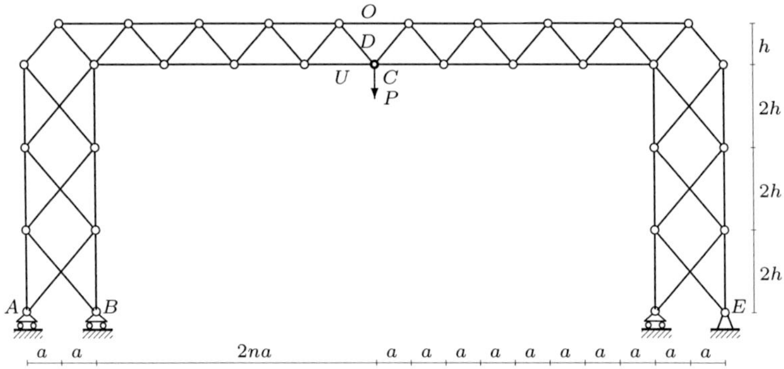

Fig. 53. Truss, $m = 3$, $n = 4$

Deflection. The coefficients in (2.11):

$C_1 = n(1 + 8n^2)/3$, $C_2 = n$, $C_3 = m$.

Support offset. The coefficients in (2.12):

$A_1 = -2n^2(1 + 2(-1)^m)$, $A_2 = 0$,

$A_3 = (2m^2 + 2(1 + (-1)^m)m - (-1)^m + 1)/2$.

Forces:

$O = -Pan/h$, $D = Pc/(2h)$, $U = Pa(2n - 1)/(2h)$.

Supports reactions:

$Y_A = 0$, $Y_B = P/2$, $X_E = 0$.

2.6.2. Load on the upper belt

Deflection. The coefficients in (2.11):

$C_1 = 10n^2(n^2 - 1)/3$, $C_2 = n^2$, $C_3 = 2m(n + 1)$.

Support offset. The coefficients in (2.12):

$A_1 = 2(-4(1 + 2(-1)^m)n^3 + (7 + 11(-1)^m)n + 3 + 3(-1)^m)/3$,

$A_2 = 1 + (-1)^m$,

$A_3 = (n + 1)(2m^2 + 2(1 + (-1)^m)m - (-1)^m + 1)$.

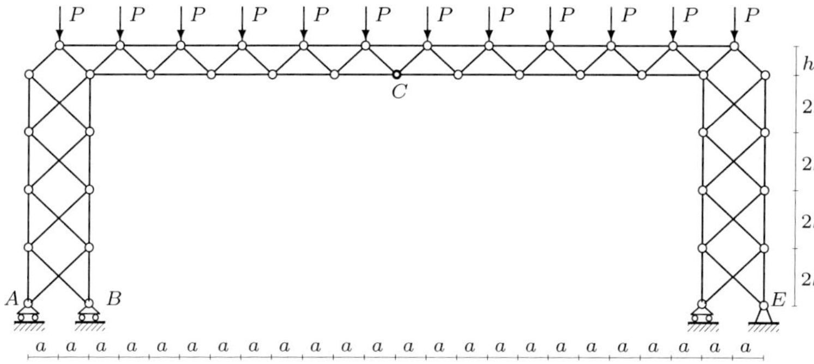

Fig. 54. Truss, $m = 4$, $n = 5$

Forces:

$O = -U = -Pa(n^2 - 1)/h$, $D = 0$.

Supports reactions: $Y_A = 0$, $Y_B = P(n + 1)$, $X_E = 0$.

Truss 2.7

A truss with a height of $(m + 1/2)h$ (Fig. 55) containing $2(n + m + 2)$ panels, consists of $8(m + n) + 4$ rods.

Thc total length of the rods in the truss is

$$L_{sum} = 2(2n + 1)a + 2(2m + 2n + 1)c + 2(2m - 1)h,$$

where $c = \sqrt{a^2 + h^2}$.

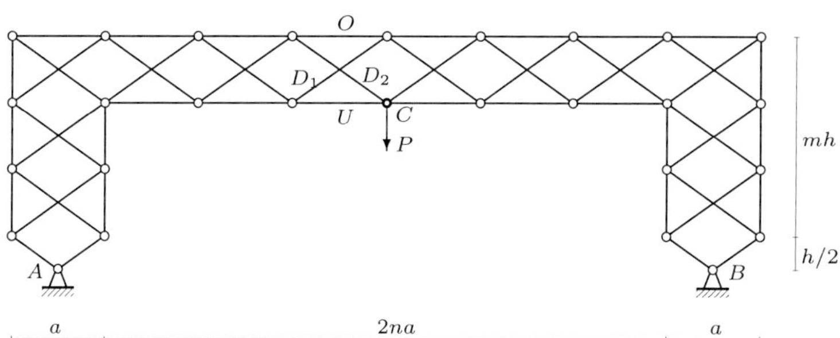

Fig. 55. Truss, $m = 3$, $n = 3$

The deflection (vertical displacement of the middle node C in the lower zone) has the form

$$\Delta = P(C_1a^3 + C_2c^3 + C_3h^3)/(h^2EF). \tag{2.13}$$

2.7.1. Concentrated force in the mid-span

Deflection. The coefficients in (2.13):

$$C_1 = (2n^3 + 3(2m((-1)^m - 1) + 2(-1)^n m + 1)n^2+$$
$$+(6(3 - 2(-1)^n - 2(-1)^m + 2(-1)^{n+m})m^2+$$
$$+6((-1)^m + (-1)^n - 1)m - 3(-1)^m + 3(-1)^{n+m} + 7-$$
$$-3(-1)^n)n + 3(2(-1)^{n+m} - 2(-1)^m + 3 - 2(-1)^n)m^2+$$
$$+3(2(-1)^m + (-1)^n - (-1)^{n+m} - 2)m - 3(-1)^n+$$
$$+3(-1)^{n+m}/2 - 3(-1)^m + 9/2)/6,$$
$$C_2 = (2(2(-1)^{n+m} - 2(-1)^n + 3 - 2(-1)^m)m^2+$$
$$+2m(-1)^m - (-1)^{n+m} + 2n)/4,$$
$$C_3 = m(2(2(-1)^{n+m} - 2(-1)^m + 3 - 2(-1)^n)m^2+$$
$$+(-1)^n - (-1)^{n+m} + 4(-1)^m)/6.$$

Case $m = n$:

$$C_1 = (2(13 - 6(-1)^n)n^3 + 12n^2 + (1 + 3(-1)^n)n - 6(-1)^n + 6)/6,$$
$$C_2 = (2(5 - 4(-1)^n)n^2 + 2(1 + (-1)^n)n - 1)/4,$$
$$C_3 = n(2(5 - 4(-1)^n)n^2 + 5(-1)^n - 1)/6.$$

Forces:

$$O = -Pa(2((-1)^m + (-1)^n - 1)m - (-1)^n + 2 + 2n - (-1)^m)/(4h),$$
$$D_1 = 0,\ D_2 = Pc/(2h),$$
$$U = Pa(2((-1)^m + (-1)^n - 1)m + (-1)^m + (-1)^n + 2n - 2)/(4h).$$

Supports reactions:

$$Y_A = Y_B = P/2,$$
$$X_A = X_B = Pa((-1)^m + (-1)^n - 1)/(2h).$$

2.7.2. Load on the upper belt

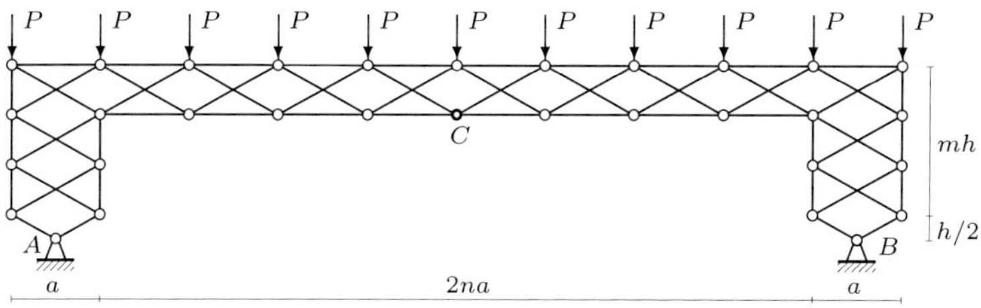

Fig. 56. Truss, $m = 3$, $n = 4$

Deflection. The coefficients in (2.13):

$$C_1 = (10n^4 + 4(5 + 2(5(-1)^m + 2(-1)^n - 5)m)n^3+$$
$$+(48(2 - 2(-1)^m - (-1)^n + (-1)^{n+m})m^2 + 12(2(-1)^n+$$
$$+7(-1)^m - 6)m + 12(-1)^{n+m} - 12(-1)^n - 24(-1)^m + 26)n^2+$$
$$+(24(4(-1)^{n+m} - 3(-1)^n + 7 - 7(-1)^m)m^2+$$
$$+4(17(-1)^m - (-1)^n - 14 - 3(-1)^{n+m})m + 34 - 12(-1)^n+$$
$$+24(-1)^{n+m} - 42(-1)^m)n + 12(-2(-1)^n + 3(-1)^{n+m}+$$
$$+5 - 5(-1)^m)m^2 + 6(7(-1)^m - 3(-1)^{n+m} - 7 - (-1)^n)m+$$
$$+9 - 15(-1)^m - 3(-1)^n + 9(-1)^{n+m})/24,$$
$$C_2 = (4(-2(-1)^n n - 4(-1)^m n + 4n - 5(-1)^m + 2(-1)^{n+m} n-$$
$$-2(-1)^n + 5 + 3(-1)^{n+m})m^2 + 2(4(-1)^m n-$$
$$-3(-1)^n + 2(-1)^m + 1 - 2n)m + 2n^2 - 2(-1)^{n+m} n-$$
$$-3(-1)^{n+m} + 3(-1)^m + (-1)^n - 5)/8,$$
$$C_3 = m(4(2((-1)^{n+m} - (-1)^n - 2(-1)^m + 2)n + 3(-1)^{n+m}-$$
$$-2(-1)^n + 5 - 5(-1)^m)m^2 + (2(8(-1)^m + (-1)^n-$$
$$-(-1)^{n+m} - 2)n - 3(-1)^{n+m} + 23(-1)^m - 5 + 5(-1)^n))/12.$$

Case $m = n$:

$$C_1 = ((114 - 88(-1)^n)n^4 + 4(53 - 33(-1)^n)n^3 + (66 - 56(-1)^n)n^2-$$

$$-2(1+9(-1)^n)n+18-18(-1)^n)/24,$$
$$C_2=(12(1-(-1)^n)n^3+5(3-2(-1)^n)n^2-(-1)^n n-4+2(-1)^n)/4,$$
$$C_3=n(12(1-(-1)^n)n^3+2(8-7(-1)^n)n^2+3(3(-1)^n-1)n+$$
$$+14(-1)^n-4)/6.$$

Forces:

$$O=-Pa(2n^2-2((-1)^m(1-2m)+2m-2)n-$$
$$-4m-1-3(-1)^m(1-2m))/(4h),$$
$$D_1=-Pc/(2h),\ D_2=0,$$
$$U=Pa(2n^2+2((-1)^m(1+2m)-2m)n+$$
$$+3(-1)^m(1+2m)-3-4m)/(4h).$$

Supports reactions: $Y_A=Y_B=P(2n+3)/2$,

$$X_B=X_A=Pa((-1)^m(2n+3)-2n-2)/(2h).$$

Truss 2.8

A truss with a height of $(m+1)h$ (Fig. 57) containing $2(n+m+2)$ panels, consists of $8(m+n)+14$ rods.

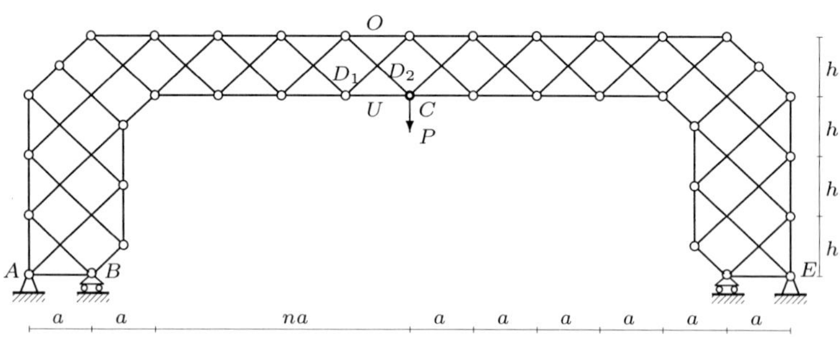

Fig. 57. Truss, $m=3$, $n=4$

The truss is unchangeable when $m=2,3,5,6,8,...$ or when $m=(6k+1-(-1)^k)/4$, $k=1,2,...$. With $m=1,4,7,10...$ the truss is kinematically variable (Fig. 58).

The total length of the rods in the truss is

$$L_{sum} = 4(n+1)a + (19 - 3(-1)^k + 8n + 18k)c/2 + (6k - 1 - (-1)^k)h,$$

where $c = \sqrt{a^2 + h^2}$.

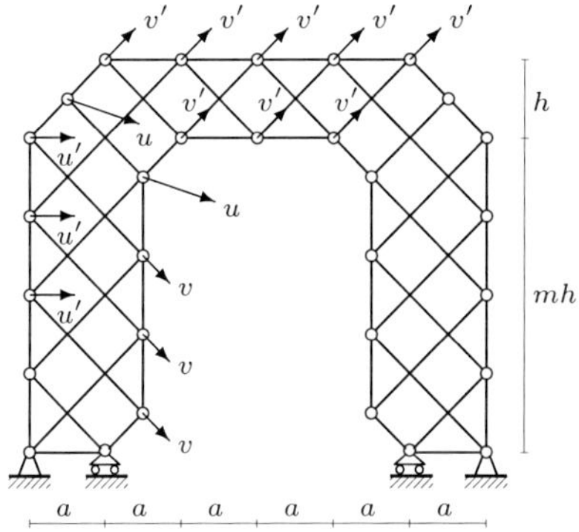

Fig. 58. Diagram of possible velocity of nodes for an instantly changeable truss, $m = 4$, $n = 1$, $u' = 2hv/c$, $c = \sqrt{a^2 + h^2}$, $u = u'd/(2h)$, $v' = 2uha/(dc)$, $d = \sqrt{a^2 + 9h^2}$.

The deflection (vertical displacement of the middle node C in the lower zone) has the form

$$\Delta = P(C_1 a^3 + C_2 c^3 + C_3 h^3)/(h^2 EF). \qquad (2.14)$$

2.8.1. Concentrated force in the mid-span

Deflection. The coefficients in (2.14):

$$C_1 = (16n^3 + 24((-1)^{k+n} + 2 - 2k)n^2 +$$
$$+16(5 + 3k^2 + 3(-1)^{k+n} - 3k(-1)^{k+n} - 6k)n + 30k^2 +$$
$$+6(3(-1)^k - 9 - 4(-1)^n - 6(-1)^{k+n})k + 69 +$$
$$+30(-1)^{k+n} - 6(-1)^n + 3(-1)^k)/48,$$

$$C_2 = (6k^3 - ((-1)^k + 1 + 6(-1)^{k+n})k^2 +$$
$$+((-1)^k - 2(-1)^{k+n} + 3 - 6(-1)^n)k + 8n +$$
$$+12 - (-1)^n + 5(-1)^{k+n})/16,$$

$$C_3 = (2k^3 - 2(1 + (-1)^{k+n})k^2 + (2(-1)^{k+n} + (-1)^k -$$
$$-2(-1)^n + 3)k + (-1)^{k+n} - (-1)^k + 1 - (-1)^n)/4.$$

Forces:

$$O = -Pa((-1)^{k+n} + 3 + 2n - 2k)/(4h),$$
$$D_1 = 0,\ D_2 = Pc/(2h),$$
$$U = Pa(2n - 2k - 1 + (-1)^{k+n})/(4h).$$

Supports reactions:

$$X_A = X_E = Pa/(2h),$$
$$Y_A = Y_E = P(3 + 2(-1)^{k+n} - (-1)^k + 2k)/8,$$
$$Y_B = P(1 - 2(-1)^{k+n} + (-1)^k - 2k)/8.$$

2.8.2. Load on the upper belt

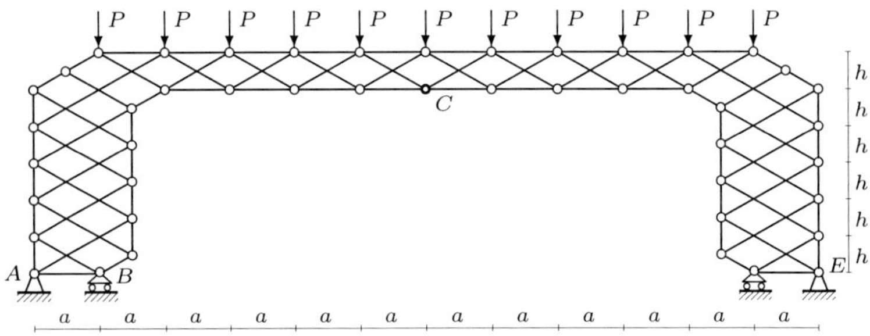

Fig. 59. Truss, $m = 3$, $n = 4$

Deflection. The coefficients in (2.14):

$$C_1 = (6(2n + 3)(8n + 5)k^2 + (-80n^3 - 24(2(-1)^{k+n} + 11)n^2 +$$
$$+(-24(-1)^n + 12(-1)^k - 292 - 108(-1)^{k+n})n + 36(-1)^k -$$
$$-126 - 36(-1)^n - 54(-1)^{k+n})k + 20n^4 + 16((-1)^{k+n} + 5)n^3 +$$
$$+4(12(-1)^{k+n} + 3(-1)^k + 34)n^2 + (6(-1)^n + 6(-1)^k + 154 +$$
$$+50(-1)^{k+n})n + 84 + 33(-1)^{k+n} - 12(-1)^k - 9(-1)^n)/48,$$
$$C_2 = (12(2n + 3)k^3 - 2(2(1 + 3(-1)^{k+n} + (-1)^k)n +$$

$$+9(-1)^{k+n}+3+6(-1)^k)k^2+2(2((-1)^k-3-3(-1)^n-$$
$$-(-1)^{k+n})n-3(-1)^n+2(-1)^k-3(-1)^{k+n}-4)k+8n^2+$$
$$+2(5(-1)^{k+n}-(-1)^n+14-2(-1)^k)n+$$
$$+21+9(-1)^{k+n}-(-1)^k+3(-1)^n)/32,$$

$$C_3=(4(3+2n)k^3-2(3(-1)^{k+n}+6+4n+(-1)^k+2(-1)^{k+n}n)k^2+$$
$$+2((2(-1)^k+2+2(-1)^{k+n}-2(-1)^n)n+4(-1)^k+2-$$
$$-(-1)^n+3(-1)^{k+n})k+2(2-2(-1)^k-(-1)^n+(-1)^{k+n})n+$$
$$+5+3(-1)^{k+n}-5(-1)^k-3(-1)^n)/8.$$

Forces:

$$O=-Pa((-1)^k+3+6n+2n^2-4nk-6k)/(4h),$$

$$D_1=-Pc/(2h),\ D_2=0,$$

$$U=Pa((-1)^k+2n-1+2n^2-4nk-6k)/(4h).$$

Supports reactions:

$$X_A=X_E=Pa(2n+3)/(2h),$$

$$Y_A=Y_E=P(6n+9+4nk+6k-(-1)^k(2n+1))/8,$$

$$Y_B=P(2n(1-2k)-6k+3+(-1)^k(2n+1))/8.$$

Truss 2.9

A truss with a height of $(2m+1)h$ (Fig. 60) containing $2(n+m+2)$ panels, consists of $8(m+n)+13$ rods.

The truss is unchangeable when $m=3,6,9,...$ or when $m=3k$, $k=1,2,...$. With $m=1,2,4,5,7,8...$ the truss is kinematically variable (Fig. 61).

The total length of the rods in the truss is

$$L_{sum}=2a(4n+5)+4h(6k-1)+2c(2n+18k+5).$$

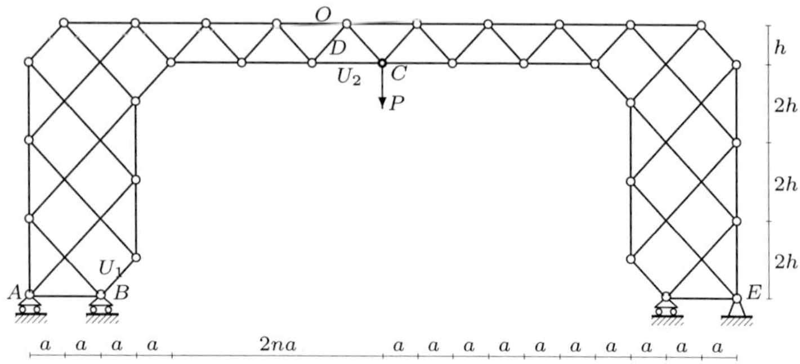

Fig. 60. Truss, $m = n = 3$

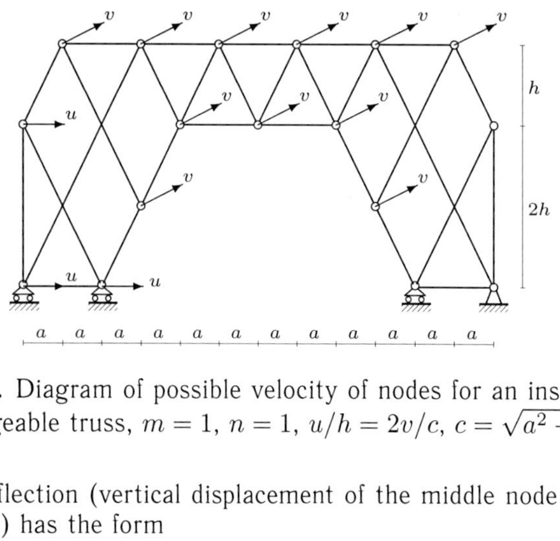

Fig. 61. Diagram of possible velocity of nodes for an instantly changeable truss, $m = 1$, $n = 1$, $u/h = 2v/c$, $c = \sqrt{a^2 + h^2}$

The deflection (vertical displacement of the middle node C in the lower zone) has the form

$$\Delta = P(C_1a^3 + C_2c^3 + C_3h^3)/(h^2EF). \tag{2.15}$$

Support A offset:

$$\delta_A = P(A_1a^3 + A_2c^3 + A_3h^3)/(ahEF). \tag{2.16}$$

2.9.1. Concentrated force in the mid-span

Deflection. The coefficients in (2.15):

$C_1 = (32n^3 + 48n^2 + 28n + 21)/12,$

$C_2 = (4n + (18k + 5))/4,\ C_3 = 7k.$

Support offset. The coefficients in (2.16):

$A_1 = (4(2n + 1)^2 k + 1 - 4n - 4n^2)/2,$

$A_2 = (18k^2 + 9k + 1)/2,\ A_3 = 2k(2 + 5k).$

Forces:

$O = -P(2n + 1)a/(2h),\ D = Pc/(2h),$

$U_1 = -Pc/(2h),\ U_2 = Pna/h.$

Supports reactions:

$X_E = 0,\ Y_A = Y_E = -P/4,\ Y_B = 3P/4.$

2.9.2. Load on the upper belt

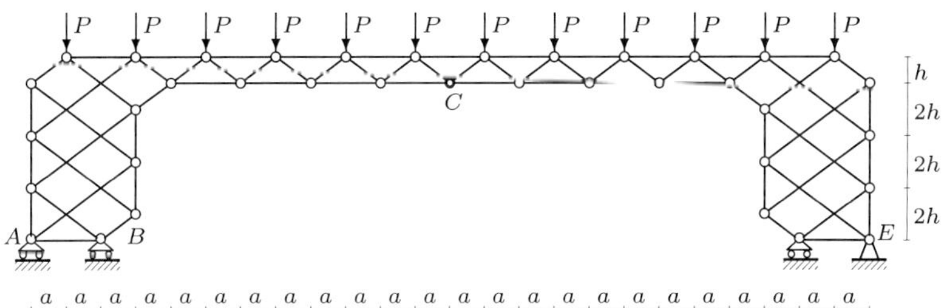

Fig. 62. Truss, $m = 3$, $n = 4$

Deflection. The coefficients in (2.15):

$C_1 = (20n^4 + 40n^3 + 28n^2 + 23n + 12)/6,$

$C_2 = (2n^2 + n(18k + 5) + 18k + 3)/2,$

$C_3 = 2(7n + 8)k.$

Support offset. The coefficients in (2.16):

$A_1 = (16n(n + 1)(2n + 1)k - 8n^3 - 12n^2 + 5n + 12)/3,$

$A_2 = 6(3n + 4)k^2 + 3(3n + 2)k + n + 1,$

$A_3 = 4k((5n + 6)k + 2n + 2).$

Forces:

$O = -Pan(n+1)/h$, $D = 0$,

$U_1 = -Pc(n+1)/h$, $U_2 = Pan(n+1)/h$.

Supports reactions: $X_E = 0$, $Y_A = Y_E = -Pn/2$,

$Y_B = P(4+3n)/2$.

Truss 2.10

A truss with a height of $(m+3/2)h$ (Fig. 63) containing $2(n+m)$ panels, consists of $8(m+n)+14$ rods. The case $m = 2$ is considered.
The deflection (vertical displacement of the middle node C in the lower zone) has the form

$$\Delta = P(C_1a^3 + C_2c^3 + C_3h^3)/(h^2EF), \quad (2.17)$$

where $c = \sqrt{a^2+h^2}$.

The total length of the rods in the truss is

$$L_{sum} = 2(2n+1)a + 2(2n+9)c + 8h.$$

2.10.1. Concentrated force in the mid-span

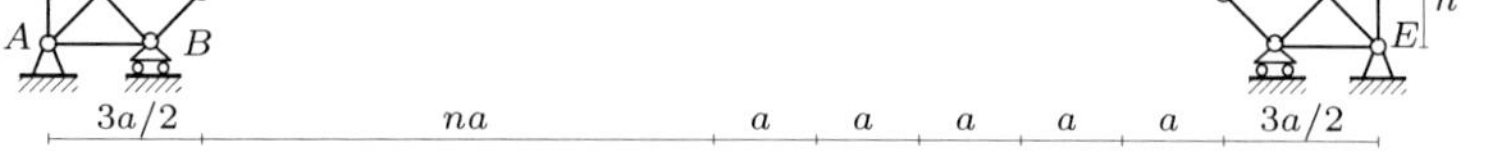

Fig. 63. Truss, $m = 2$, $n = 5$

Deflection. The coefficients in (2.17):

$C_1 = (4n^3 + 6(1-2(-1)^n)n^2 + 4(5-3(-1)^n)n + 3 + 3(-1)^n)/12$,

$C_2 = (16 - 11(-1)^n + 2n)/4$, $C_3 = 5 - 3(-1)^n$.

Forces:

$O = -Pa(2n+1-(-1)^n)/(4h)$, $D_1 = 0$, $D_2 = Pc/(2h)$,

$U = Pa(2n-1-3(-1)^n)/(4h)$.

Supports reactions:

$X_A = X_E = P(-1)^n a/(2h)$, $Y_A = Y_E = P(-1)^n$,

$Y_B = P(1-2(-1)^n)/2$.

2.10.2. Load on the upper belt

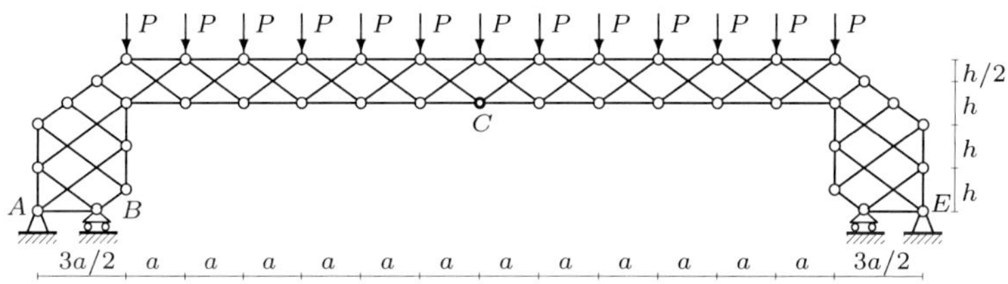

Fig. 64. Truss, $m = 2$, $n = 6$

Deflection. The coefficients in (2.17):

$$C_1 = (5n^4 + 2(5-4(-1)^n)n^3 + (13-12(-1)^n)n^2 + \\ +2(4-11(-1)^n)n - 3(-1)^n - 3)/12,$$

$$C_2 = (n^2 + (6-11(-1)^n)n + 8 - 18(-1)^n)/4,$$

$$C_3 = 3(1-(-1)^n)n + 3 - 5(-1)^n.$$

Forces:

$O = -Pa(n+1)n/(2h)$, $D_1 = -Pc/(2h)$,

$D_2 = 0$, $U = Pa/(2h)$.

Supports reactions:

$X_A = X_E = Pa/(2h)$, $Y_A = Y_E = -P$, $Y_B = P(3+2n)/2$.

Truss 2.11

A truss with a height of $(m+1/2)h$ (Fig. 65) containing $2(n+m)$ panels, consists of $8(m+n)+8$ rods. The total length of the rods in the truss is

$$L_{sum} = 4(n+1)a + (6m+4n+3)c + (4m-2)h,$$

where $c = \sqrt{a^2 + h^2}$.

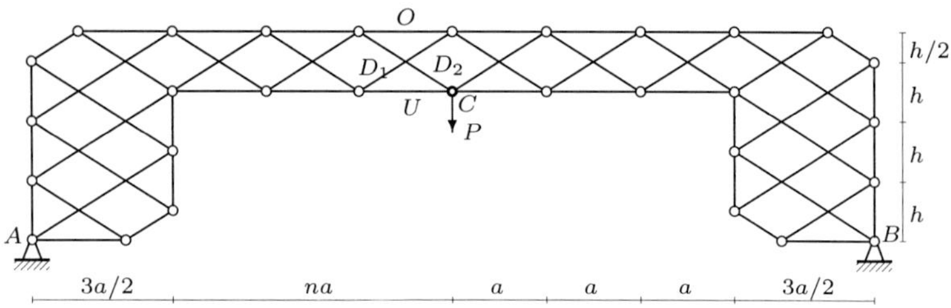

Fig. 65. Truss, $m = 3$, $n = 3$

The truss is unchangeable when $m = 2, 3, 5, 6, 8, ...$ or when $m = (6k + 1 - (-1)^k)/4$, $k = 1, 2, ...$. With $m = 1, 4, 7, 11, ...$ the truss is kinematically variable (Fig. 66).

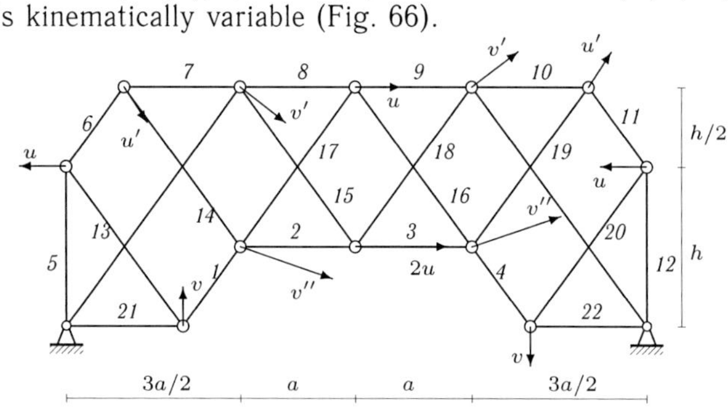

Fig. 66. Diagram of possible velocity of nodes for an instantly changeable truss, $m = n = 1$, $\omega_k = \omega$, $k = 2, 3, 8, 9, 10, 12, .., 22$, $\omega_k = -4\omega$, $k = 1, 4, 6, 11$, $\omega_{23} = \omega_{24} = -2\omega/3$

The case $m = 3$ is considered.

The deflection (vertical displacement of the middle node C in the lower zone) has the form

$$\Delta = P(C_1 a^3 + C_2 c^3 + C_3 h^3)/(h^2 EF). \tag{2.18}$$

2.11.1. Concentrated force in the mid-span

Deflection. The coefficients in (2.18):

$$C_1 = (12n^3 + 18(2(-1)^n - 1)n^2 + 8(7 - 5(-1)^n)n+$$
$$+85 - 77(-1)^n)/36,$$

$$C_2 = (2n + 11 - 10(-1)^n)/4,\ C_3 = (31 - 23(-1)^n)/9.$$

Forces:

$$O = -Pa(5(-1)^n + 6n - 1)/(12h),\ D_1 = 0,\ D_2 = Pc/(2h),$$
$$U = Pa(7(-1)^n + 6n - 11)/(12h).$$

Supports reactions: $X_A = Pa(2 - (-1)^n)/(6h)$, $Y_A = Y_B = P/2$.

2.11.2. Load on the upper belt

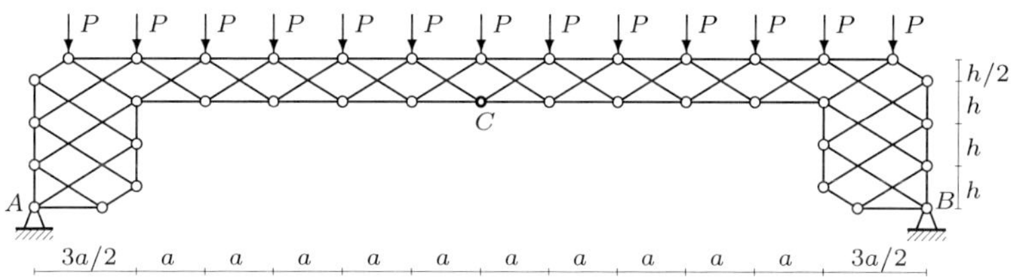

Fig. 67. Truss, $m = 3$, $n = 5$

Deflection. The coefficients in (2.18):

$$C_1 = (15n^4 + 6(4(-1)^n - 5)n^3 - 5(8(-1)^n + 5)n^2+$$
$$+2(55 - 86(-1)^n)n - 103(-1)^n + 71)/36,$$

$$C_2 = (6n^2 + 12(4 - 5(-1)^n)n + 47 - 61(-1)^n)/24,$$

$$C_3 = ((31 - 23(-1)^n)n - 19(-1)^n + 35)/9.$$

Forces:

$$O = -Pa(3n^2 - n - 8)/(6h),\ D_1 = -Pc/(2h),\ D_2 = 0,$$
$$U = Pa(3n^2 - 5n - 10)/(6h).$$

Supports reactions:

$$X_A = Pa(4n + 5)/(6h),\ Y_A = Y_B = P(2n + 3)/2.$$

Truss 2.12

A truss with a height of $(m+2)h$ (Fig. 68) containing $2n+$ $+2m$ panels, consists of $8n+8m+13$ rods. The deflection (vertical displacement of the middle node C in the lower zone) has the form

$$\Delta = P(C_1a^3 + C_2c^3 + C_3g^3 + C_4h^3)/(h^2EF), \tag{2.19}$$

where $c=\sqrt{a^2+h^2}$, $g=\sqrt{a^2+4h^2}$.

The total length of the rods in the truss is

$$L_{sum} = 2(m+2n+1)a + (2m+4n+3)c + 4(m+n+1)h + 2g.$$

Support A offset:

$$\delta_A = P(A_1a^3 + A_2c^3 + A_3h^3)/(ahEF). \tag{2.20}$$

2.12.1. Concentrated force in the mid-span

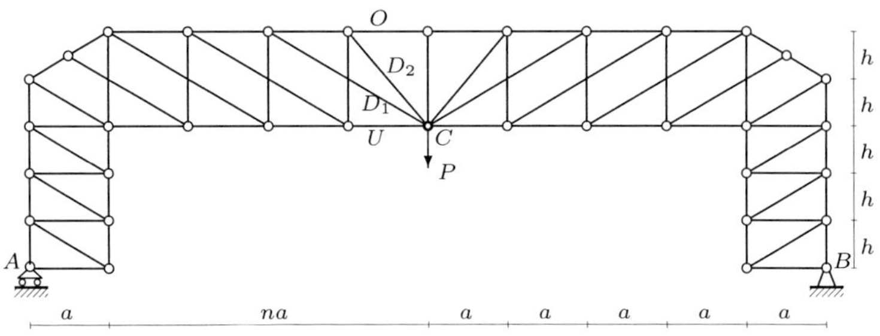

Fig. 68. Truss, $m=3$, $n=4$

Deflection. The coefficients in (2.19):

$C_1 = (12n^2 + 20n - 3 + 3(-1)^n + 4n^3)/48$, $C_2 = (2n + (-1)^n)/4$,

$C_3 = (1 - (-1)^n)/16$, $C_4 = (2n - (-1)^n + 2m)/4$.

Support offset. The coefficients in (2.20):

$A_1 = (2(m+1)n^2 + 4nm - (-1)^n + 1)/8$, $A_2 = m/2$, $A_3 = m^2/2$.

Forces:

$O = -Pa(n+1)/(4h)$, $D_1 = Pc((-1)^n + 1)/(4h)$,

$D_2 = -Pg((-1)^n - 1)/(8h)$,

$U = Pa(2n - (-1)^n - 1)/(8h)$.

Supports reactions: $Y_A = Y_B = P/2$, $X_B = 0$.

2.12.2. Load on the upper belt

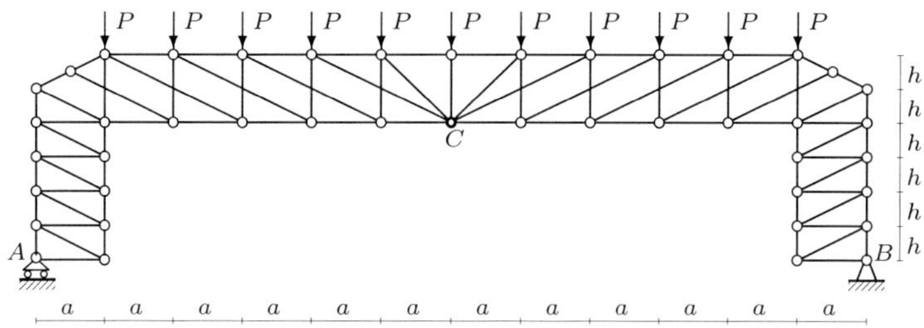

Fig. 69. Truss, $m = 4$, $n = 5$

Deflection. The coefficients in (2.19):

$C_1 = (5n^4 + 20n^3 + 40n^2 + (3(-1)^n + 7)n + 3((-1)^n - 1)/2)/48$,

$C_2 = (6n^2 + 2(1 + (-1)^n)n + 1 + (-1)^n)/8$,

$C_3 = (1 - (-1)^n)n/16$,

$C_4 = (6n^2 + 2(4m - (-1)^n + 3)n - 3(-1)^n + 4m + 1)/8$.

Support offset. The coefficients in (2.20):

$A_1 = n(4(m + 1)n^2 + 6(2m + 1)n + 5 - 3(-1)^n + 5m)/12$,

$A_2 = m(1 + 2n)/2$,

$A_3 = m^2(1 + 2n)/2$.

Forces:

$O = -Pa(n + 1)^2/(4h)$, $D_1 = Pc(1 + (-1)^n n)/(2h)$,

$D_2 = -(-1)^n Pgn/(4h)$,

$U = Pa(n^2 + 2n - (-1)^n n - 1)/(4h)$.

Supports reactions:

$Y_A = Y_B = (2n + 1)P/2$, $X_B = 0$.

Truss 2.13

A truss with a height of $(2m+3)h$ (Fig. 70) containing $2(n+m)$ panels, consists of $8(m+n)+8$ rods.

The deflection (vertical displacement of the middle node C in the lower zone) has the form

$$\Delta = P(C_1a^3 + C_2c^3 + C_3h^3)/(h^2(2m+1)^2EF), \tag{2.21}$$

where $c = \sqrt{a^2+h^2}$.

The total length of the rods in the truss is

$$L_{sum} = 8an + 2(2m+4n+3)c + 4(2m+1)h.$$

2.13.1. Concentrated force in the mid-span

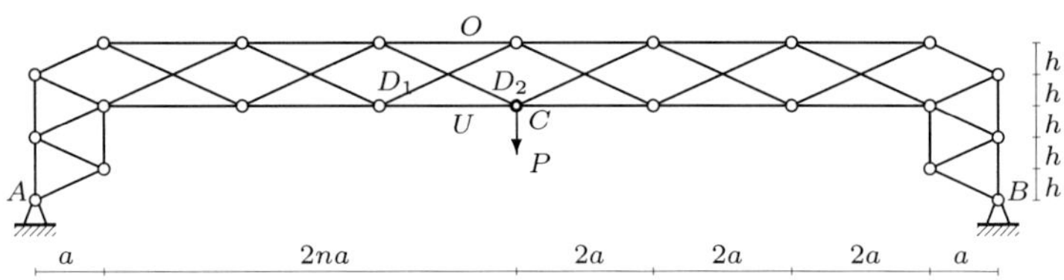

Fig. 70. Truss, $m = 1$, $n = 3$

Deflection. The coefficients in (2.21):

$$C_1 = (4(2m+1)^2n^3 + 6(2m+1)(2m-2(-1)^nm+1-2(-1)^n)n^2+$$
$$+4(2(4-3(-1)^n)m^2 + (11-9(-1)^n)m + 5 - 3(-1)^n)n+$$
$$+3(2m+1)((-1)^n-1))/6,$$

$$C_2 = (2(2m+1)^2n + (2m+1)(2-2(-1)^nm + 2m - (-1)^n))/2,$$

$$C_3 = (2(m+1)(2m+1)(5m-3(-1)^nm+3-3(-1)^n))/3.$$

Forces:

$$O = -Pa(1+2n-(-1)^n)/(4h),$$

$$D_1 = 0,\ D_2 = Pc/(2h),$$

$$U = Pa(2(2m+1)n - 2m((-1)^n+1) - 1 - 3(-1)^n)/(4h(2m+1)).$$

Supports reactions:

$Y_A = Y_B = P/2,$

$X_A = X_B = (-1)^n a/(2h(2m+1)).$

2.13.2. Load on the upper belt

Deflection. The coefficients in (2.21):

$$C_1 = (5(2m+1)^2n^4+2(2m+1)(10m-4(-1)^nm-4(-1)^n+5)n^3+ \\ +(2m+1)(20m-12(-1)^nm+13-12(-1)^n)n^2+ \\ +2(10(1-(-1)^n)m^2+(13-21(-1)^n)m-11(-1)^n+4)n+ \\ +3(m+1)(2m+1)((-1)^n-1))/6,$$

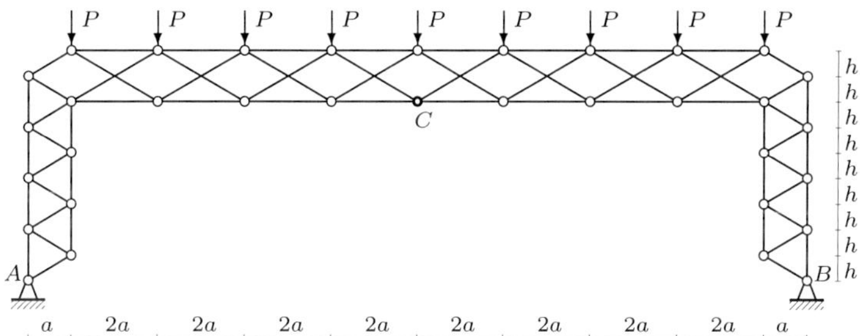

Fig. 71. Truss, $m = 3$, $n = 4$

$$C_2 = (2(2m+1)^2n^2+2(2m+1)^2(-(-1)^n+1)n+ \\ +1-3(-1)^n-4(-1)^nm^2-8(-1)^nm+4m+4m^2)/4,$$

$$C_3 = (6(m+1)(2m+1)(2m-(-1)^nm+1-(-1)^n)n+ \\ +(m+1)(2m+1)(9m+6-6(-1)^n-7(-1)^nm))/3.$$

Forces:

$O = -Pa(n+1)n/(2h)$, $D_1 = -Pc/(2h)$, $D_2 = 0$,

$U = Pa(n+n^2+2n^2m+2nm+2m+2)/(2h(2m+1)).$

Supports reactions:

$Y_A = Y_B = (2n+1)P/2$, $X_A = X_B = Pa/(2h(2m+1)).$

Truss 2.14

A truss with a height of $(m + k + 1)h$ (Fig. 72) containing $2(n+m+k)$ panels, consists of $8(m+n+k)+5$ rods. The total length of the rods in the truss is

$$L_{sum} = 2(m+2n+k+1)a+2(m+n+2k+1)c+(4m+2n+2k+1)h,$$

where $c = \sqrt{a^2 + h^2}$.

The deflection (vertical displacement of the middle node C in the lower zone) has the form

$$\Delta = P(C_1a^3 + C_2c^3 + C_3h^3)/(h^2EF). \qquad (2.22)$$

Support A offset:

$$\delta_A = P(A_1a^3 + A_2c^3 + A_3h^3)/(ahEF). \qquad (2.23)$$

2.14.1. Concentrated force in the mid-span

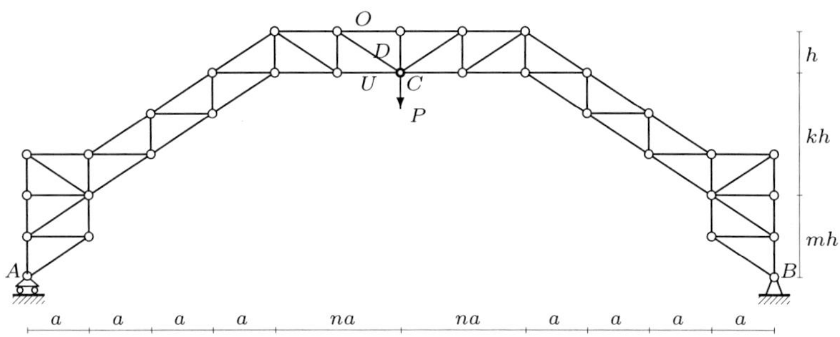

Fig. 72. Truss, $m = 2$, $n = 2$, $k = 3$

Deflection. The coefficients in (2.22):

$C_1 = (6nk^2 + 3(2n^2 + 4n + 1)k + (n + 1)(2n^2 + 4n + 3))/6,$

$C_2 = (3n + 7k + 3 + 2k^3 + 6k^2)/6,$

$C_3 = (m + 3 + n + k^2 + k)/2.$

Case $m = k = n$:

$C_1 = (14n^3 + 18n^2 + 10n + 3)/6,$

$C_2 = (2n^3 + 6n^2 + 10n + 3)/6,$

$C_3 = (n^2 + 3n + 3)/2.$

Support offset. The coefficients in (2.23):

$A_1 = (4nk^2 + 2(n^2 + (2m+3)n+1)k + (n+1)((2m+1)n+2m))/2,$

$A_2 = (4k^3 + 3(2m+3)k^2 + (12m+5)k + 6m)/6,$

$A_3 = k^2 + (m+1)k + (7m + m^2 + 2)/2.$

Case $m = k = n$:

$A_1 = n(12n^2 + 11n + 5)/2,$

$A_2 = n(10n^2 + 21n + 11)/6,$

$A_3 = (2 + 5n^2 + 9n)/2.$

Forces:

$O = -Pa(n + k + 1)/(2h)$, $D = Pc/(2h)$, $U = Pa(k + n)/(2h)$.

Supports reactions: $X_B = 0$, $Y_A = Y_B = P/2$.

2.14.2. Load on the upper belt

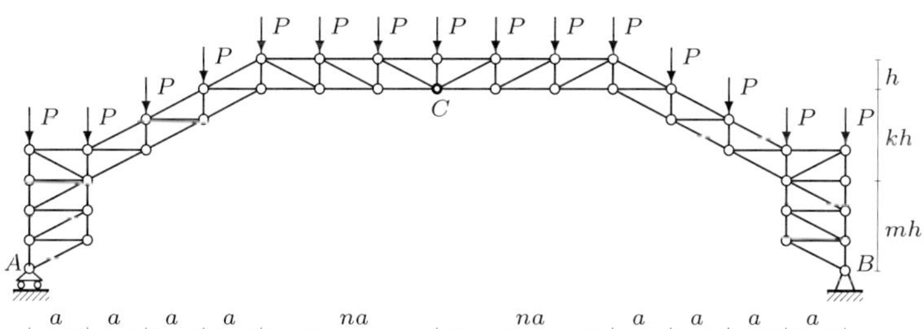

Fig. 73. Truss, $m = n = k = 3$

Deflection. The coefficients in (2.22):

$$C_1 = (12nk^3 + 6(n+1)(5n+1)k^2 + 2(n+2)(10n^2 + 10n + 3)k+$$
$$+(5n^2 + 10n + 6)(n+1)^2)/12,$$

$$C_2 = (5k^4 + 4(2n+5)k^3 + (24n+31)k^2 + 2(14n+11)k+$$
$$+6(n+1)^2)/12,$$

$C_3 = (k^3+(3+2n)k^2+2(m+5+n)k+3m+5+n^2+2(m+4)n)/2.$

Case $m = k = n$:

$C_1 = (67n^4 + 116n^3 + 83n^2 + 34n + 6)/12,$

$C_2 = (13n^4 + 44n^3 + 65n^2 + 34n + 6)/12,$

$C_3 = (3n^3 + 10n^2 + 21n + 5)/2.$

Support offset. The coefficients in (2.23):

$$A_1 = (12nk^3 + 6(4n^2 + (2m + 5)n + 1)k^2+$$
$$+2(4n^3 + 6(2m + 3)n^2 + (12m + 17)n + 6m)k+$$
$$+(n + 1)(4(2m + 1)n^2 + (16m + 5)n + 6m))/6,$$
$$A_2 = (5k^4 + 8(n + m + 2)k^3 + 2(3(2m + 3)n + 12m + 8)k^2+$$
$$+(2(12m + 5)n + 22m + 5)k + 6m + 12nm)/6,$$
$$A_3 = k^3 + (2n + 3 + m)k^2 + (2n(m + 1) + m^2 + 9m + 4)k+$$
$$+(m^2 + 7m + 2)n + (2 + 13m + 3m^2)/2.$$

Case $m = k = n$:

$A_1 = n(88n^3 + 118n^2 + 83n + 11)/6,$

$A_2 = n(33n^3+82n^2+60n+11)/6,\ A_3 = (16n^3+45n^2+25n+2)/2.$

Forces:

$O = -Pa(n + k + 1)^2/(2h),\ D = Pc/(2h),$

$U = Pa(n + k + 2)(k + n)/(2h).$

Supports reactions:

$X_B = 0,\ Y_A = Y_B = (2n + 2k + 3)P/2.$

Truss 2.15

A truss with a height of $(m+2)h$ (Fig. 74) containing $2n+2m+2$ panels, consists of $8(n + m) + 13$ rods. The total length of the rods in the truss is

$$L_{sum} = 2(m + 2n + 2)(a + c) + 2(2m + 2n + 3)h,$$

where $c = \sqrt{a^2 + h^2}$.

The deflection (vertical displacement of the middle node C in the lower zone) has the form

$$\Delta = P(C_1 a^3 + C_2 c^3 + C_3 h^3)/(h^2 EF). \quad (2.24)$$

Support A offset:

$$\delta_A = P(A_1 a^3 + A_2 c^3 + A_3 h^3)/(ahEF). \quad (2.25)$$

2.15.1. Concentrated force in the mid-span

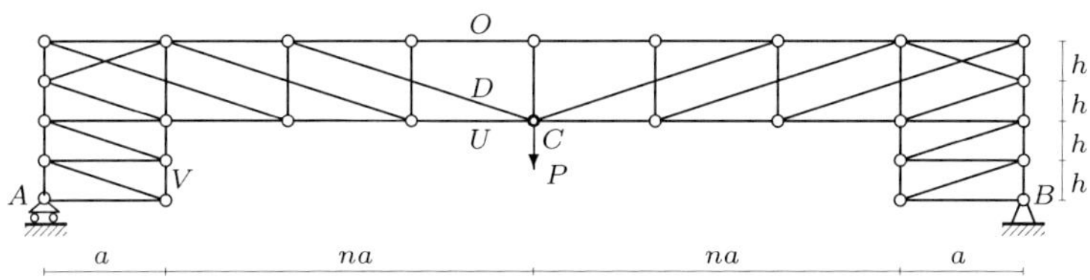

Fig. 74. Truss, $m = 2$, $n = 3$

Deflection. The coefficients in (2.24):

$C_1 = (6n^2 + 10n + 3 - 3(-1)^n + 2n^3)/24,$

$C_2 = (4n + 3 - (-1)^n)/8,\ C_3 = (4n + 1 - 3(-1)^n + 4m)/8.$

Support offset. The coefficients in (2.25):

$A_1 = (2(m + 1)n^2 + 4nm + (m + 1)((-1)^n - 1))/8,$

$A_2 = m(1 + (-1)^n)/4,\ A_3 = -m((-1)^n - 1 - 2m)/4.$

Forces:

$O = -Pa(n+1)/(4h),\ D = Pc/(2h),\ U = Pa(n-1)/(4h),\ V = 0.$

Supports reactions:

$Y_A = Y_B = P/2,\ X_B = 0.$

2.15.2. Load on the upper belt

Deflection. The coefficients in (2.24):

$C_1 = (10n^4 + 40n^3 + 68n^2 + 4(5 - 3(-1)^n)n - 3 + 3(-1)^n)/96,$

$C_2 = (2n^2 + (3 - (-1)^n)n + 2)/8,$

$$C_3 = (2n^2 + (8m - 3(-1)^n + 13)n + 10 + 12m - 4(-1)^n)/8.$$

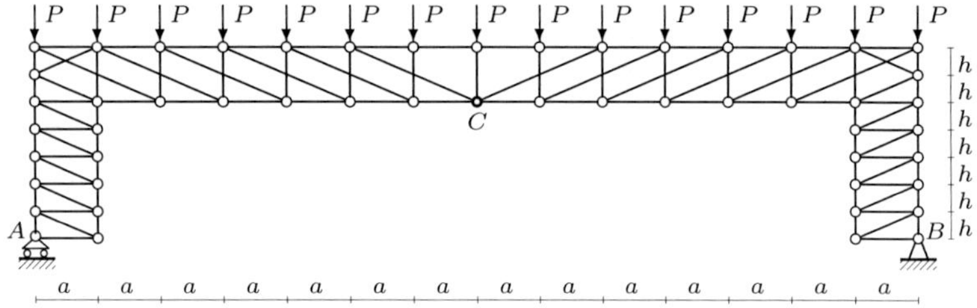

Fig. 75. Truss, $m = 5$, $n = 6$.

Support offset. The coefficients in (2.25):

$A_1 = (2(m+1)n^3 + 3(1+2m)n^2 + (m+1)n)/6,$

$A_2 = m(n+1)/2,\ A_3 = m(n+2+(2n+3)m)/2.$

Forces:

$O = -Pa(n+1)^2/(4h),\ D = Pc/(2h),$

$U = Pa(n^2+2n-1)/(4h),\ V = 0.$

Supports reactions:

$Y_A = Y_B = (2n+3)P/2,\ X_B = 0.$

Truss 2.16

A truss with a height of $(2m+1)h/2$ consists of $8(n+m)+7$ rods (Fig. 76). The total length of the rods in the truss is

$$L_{sum} = (4n+3)a + (4m+2n+1/2)c + 2(2m-1)h + d,$$

where $c = \sqrt{a^2+h^2}$, $d = \sqrt{a^2+9h^2}$.

The deflection (vertical displacement of the middle node C in the lower zone) has the form [29]

$$\Delta = P(C_1a^3 + C_2c^3 + C_3h^3 + C_4d^3)/(h^2EF). \tag{2.26}$$

Support A offset has the form

$$\delta_A = P(A_1a^3 + A_2c^3 + A_3h^3 + A_4d^3)/(ahEF). \tag{2.27}$$

2.16.1. Concentrated force in the mid-span

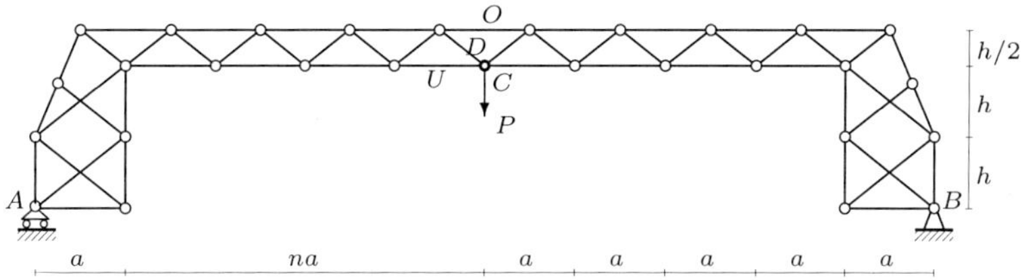

Fig. 76. Truss, $m = 2$, $n = 4$

Deflection. The coefficients in (2.26):

$C_1 = (6 + 25n + 8n^3 + 24n^2)/6$,

$C_2 = (8n + 11)/16$, $C_3 = (m - 1)/2$, $C_4 = 1/16$.

Support offset. The coefficients in (2.27):

$A_1 = (1 + 4m)n^2 + 2(1 + 4m)n + 2m$, $A_2 = (11m + 4)/8$,

$A_3 = (1 + 2m^2 - 4m - (-1)^m)/4$, $A_4 = m/8$.

Forces:

$O = -Pa(n + 1)/h$, $D = Pc/(2h)$, $U = Pa(2n + 1)/(2h)$.

Supports reactions: $Y_A = Y_B = P/2$, $X_B = 0$.

2.16.2. Load on the upper belt

Deflection. The coefficients in (2.26):

$C_1 = (n + 1)(5n^3 + 15n^2 + 16n + 3)/3$,

$C_2 = (4n^2 + 11n + 5)/8$, $C_3 = (n + 1)(m - 1)$, $C_4 = (n + 1)/8$.

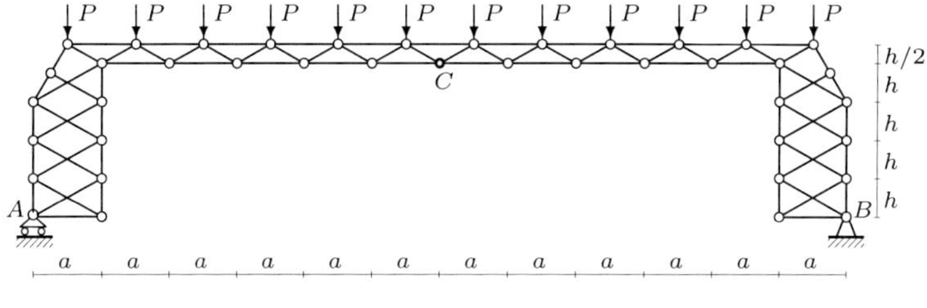

Fig. 77. Truss, $m = 4$, $n = 5$.

Support offset. The coefficients in (2.27):

$A_1 = 2(n+1)(2n^2 + 8n^2m + 16nm + 4n + 3m)/3,$

$A_2 = ((11m+4)n + 5m + 4)/4,$

$A_3 = (n+1)(2m^2 - 4m + 1 - (-1)^m)/2,\ A_4 = m(n+1)/4.$

Forces:

$O = -Pa(n+1)^2/h,\ D = 0,\ U = Pa(n+1)^2/h.$

Supports reactions: $Y_A = Y_B = (n+1)P,\ X_B = 0.$

Truss 2.17

A truss with a height of $(m+1)h$ consists of $8n + 8m - 1$ rods (Fig. 78). The total length of the rods in the truss is

$$L_{sum} = 4(m + 2n - 1)a + 2(2m + 2n - 1)c + 4hm,$$

where $c = \sqrt{a^2 + h^2}$.

The deflection (vertical displacement of the middle node C in the upper zone) has the form

$$\Delta = P(C_1a^3 + C_2c^3 + C_3h^3)/(h^2EF). \tag{2.28}$$

Support A offset:

$$\delta_A = P(A_1a^3 + A_2c^3 + A_3h^3)/(ahEF). \tag{2.29}$$

2.17.1. Concentrated force in the mid-span

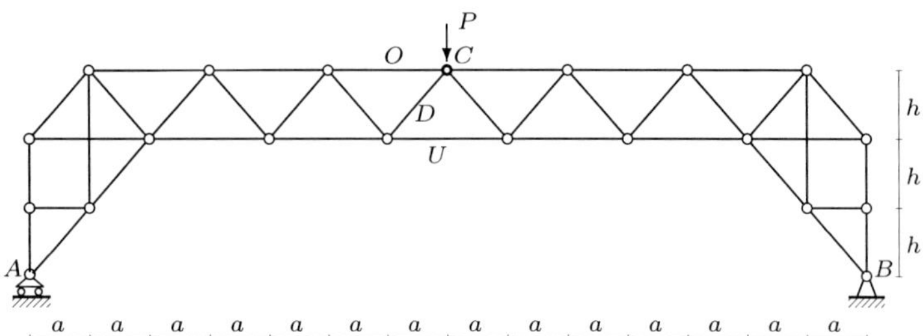

Fig. 78. Truss, $m = 2$, $n = 3$

Deflection. The coefficients in (2.28):

$$C_1 = (6(4n-1)m^2 + 6(8n^2 - 8n + 3)m +$$
$$+3(-1)^m + 32n^3 - 48n^2 + 28n - 9)/12,$$

$$C_2 = (2m^3 + (7 - 3(-1)^m)m - 12 + 24n)/24,$$

$$C_3 = (2m + 1 - (-1)^m)/4.$$

Support offset. The coefficients in (2.29):

$$A_1 = (2(4n-1)m^2 + 2(4n^2 - 2n + 1)m + (-1)^m + 4n^2 - 4n + 1)/2,$$

$$A_2 = (4m^3 + 6m^2 + 2(1 - 3(-1)^m)m + 3 - 3(-1)^m)/24,$$

$$A_3 = (2m + 1 - (-1)^m)/2.$$

Forces:

$$O = -Pa(m - 2 + 2n)/(2h),$$

$$D = -Pc/(2h),\ U = Pa(m - 1 + 2n)/(2h).$$

Supports reactions: $Y_A = Y_B = P/2$, $X_B = 0$.

2.17.2. Load on the upper belt

Deflection. The coefficients in (2.28):

$$C_1 = (6(2n+1)(4n-1)m^2 + 2(10n-9)(4n^2-1)m +$$
$$+40n^4 - 9 + 14n - 80n^3 + 20n^2 + 6(-1)^m n + 3(-1)^m)/12,$$

$$C_2 = (2(2n+1)m^3 + (14n - 5 - 3(-1)^m - 6(-1)^m n)m +$$
$$+6 + 6(-1)^m + 24n^2 - 24n)/24,$$

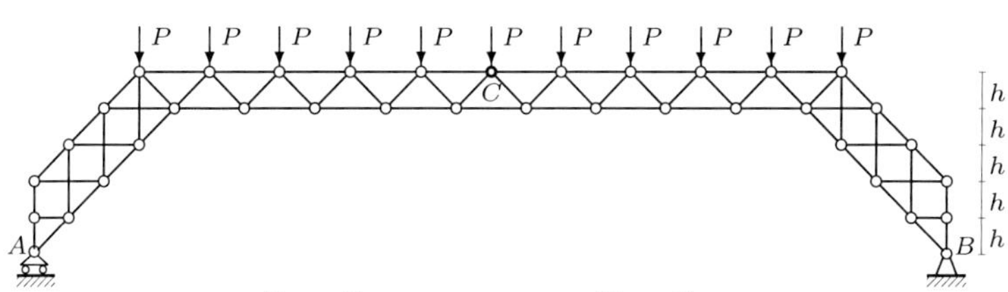

Fig. 79. Truss, $m = 4$, $n = 5$

$C_3 = (2n+1)(1+2m-(-1)^m)/4.$

Support offset. The coefficients in (2.29):

$$A_1 = (2n+1)(6(4n-1)m^2 + 2(2n-1)(4n-3)m+$$
$$+3(-1)^m + 8n^2 - 16n + 3)/6,$$
$$A_2 = (4(2n+1)m^3 + 6(2n+1)m^2+$$
$$+(4(1-3(-1)^m)n - 6(-1)^m - 22)m-$$
$$-3(2n-3)((-1)^m - 1))/24,$$
$$A_3 = (2(2n+1)m - 2(-1)^m n - (-1)^m + 1 + 2n)/2.$$

Forces: $O = -Pa(m - 2 + 2nm - 2n + 2n^2)/(2h)$,

$D = -Pc/(2h)$, $U = Pa(2nm + m - 1 - 2n + 2n^2)/(2h)$.

Supports reactions: $Y_A = Y_B = P(2n+1)/2$, $X_B = 0$.

Truss 2.18

A truss with a height of $(m+2)h$ consists of $8m + 8n + 13$ rods (Fig. 80). The total length of the rods in the truss is

$$L_{sum} = 2a(m+2+2n) + 2c(n+m+1) + 2d(n+1) + h(4m+5).$$

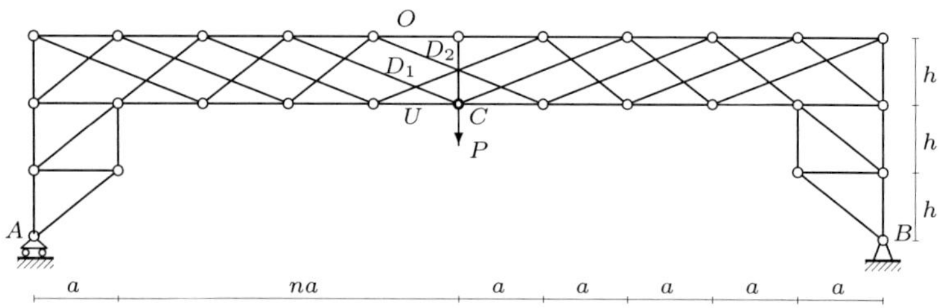

Fig. 80. Truss, $m = 1$, $n = 4$

The deflection (vertical displacement of the middle node C in the lower zone) has the form

$$\Delta = P(C_1a^3 + C_2c^3 + C_3h^3 + C_4d^3)/(h^2EF), \qquad (2.30)$$

where $c = \sqrt{a^2+h^2}$, $d = \sqrt{4a^2+h^2}$.

Support A offset:

$$\delta_A = P(A_1a^3 + A_2c^3 + A_3h^3 + A_4d^3)/(ahEF). \tag{2.31}$$

2.18.1. Concentrated force in the mid-span

The truss is unchangeable when $n = 1, 2, 4, 5, 7, \ldots$ or when $n = (6k - (-1)^k - 3)/4$, $k = 1, 2, \ldots$. With $n = 3k$ the truss is kinematically variable.

Deflection. The coefficients in (2.30):

$C_1 = (18k^3 + 9(1 - (-1)^k)k^2 + 3(5 - (-1)^k)k + 5 - 5(-1)^k)/16,$

$C_2 = (2k + (-1)^k - 1)/8,\ C_3 = (2m + 3 - (-1)^k)/4,$

$C_4 = (2k + 1 - (-1)^k)/8.$

Support offset. The coefficients in (2.31):

$A_1 = (18(2m + 3)k^2 + 6(1 - 3(-1)^k - 2(-1)^k m + 2m)k-$

$-31 - 49(-1)^k - 50(-1)^k m - 14m)/16,$

$A_2 = 0,\ A_3 = (m + 1)(m + 2)/2,\ A_4 = 0.$

Forces:

$O = -Pa(1 - (-1)^k + 6k)/(8h),\ D_1 = Pd/(2h),\ D_2 = 0,$

$U = Pa(6k - 7 - (-1)^k)/(8h).$

Supports reactions:

$Y_A = Y_B = P/2,\ X_B = 0.$

2.18.2. Load on the upper belt

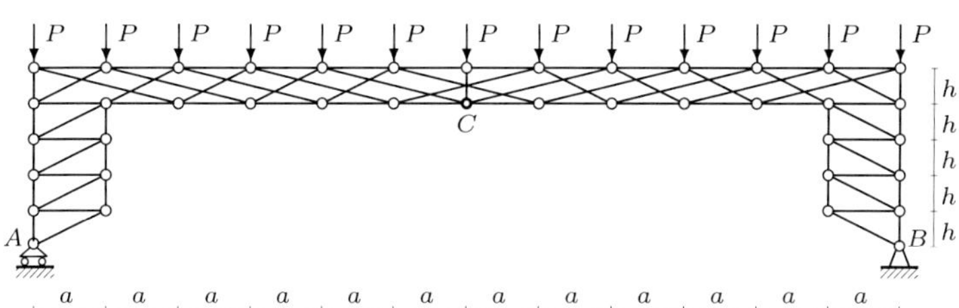

Fig. 81. Truss, $m = 3$, $n = 5$

Deflection. The coefficients in (2.30):

$C_1 = (270k^4 + 180(1-(-1)^k)k^3 + 18(1-5(-1)^k)k^2 +$
$+4(15-47(-1)^k)k + 109(-1)^k - 109)/128,$

$C_2 = (2k^2 + (6+2(-1)^k)k + 3(-1)^k - 3)/16,$

$C_3 = ((7+6m-(-1)^k)k + 3m - 3(-1)^k + 5 - (-1)^k m)/4,$

$C_4 = (2k^2 + 2(1-(-1)^k)k - (-1)^k + 1)/16.$

Support offset. The coefficients in (2.31):

$A_1 = (36(2m+3)k^3 + 6((22-6(-1)^k)m - 9(-1)^k + 22)k^2 -$
$-4((7(-1)^k+3)m + 3(-1)^k + 4)k - 7(-1)^k - 64m - 73)/16,$

$A_2 = (2k^2 - 2(1+(-1)^k)k - (-1)^k + 1)(m+1)/2,$

$A_3 = (m+1)((6m+16+4(-1)^k)k - (-1)^k m - 2(-1)^k + 3m + 6)/4,$

$A_4 = (m+1)(2k^2 - 1 + (-1)^k)/2.$

Forces:

$O = -Pa(18k^2 + (6-22(-1)^k)k - 7 + 7(-1)^k)/(16h),$

$D_1 = Pd/2/h,$

$D_2 = Pd((-1)^k - 1 - 2(-1)^k k)/(4h),$

$U = Pa(18k^2 + 2(3+5(-1)^k)k - 9(-1)^k - 7)/(16h).$

Supports reactions:

$Y_A = Y_B = P(2n+3)/2,\ X_B = 0.$

Truss 2.19

A truss with a height of $(m+1)h$ consists of $8m+8n+13$ rods (Fig. 82).The deflection (vertical displacement of the middle node C in the lower zone) has the form

$$\Delta = P(C_1a^3 + C_2c^3 + C_3h^3)/(h^2EF), \tag{2.32}$$

where $c = \sqrt{a^2+h^2}$.

The total length of the rods in the truss is

$$L_{sum} = (3 + 4n)a + (4m + 4n + 7)c + 2(1 + 2m)h.$$

Support A offset:

$$\delta_A = P(A_1 a^3 + A_2 c^3 + A_3 h^3)/(ahEF). \tag{2.33}$$

2.19.1. Concentrated force in the mid-span

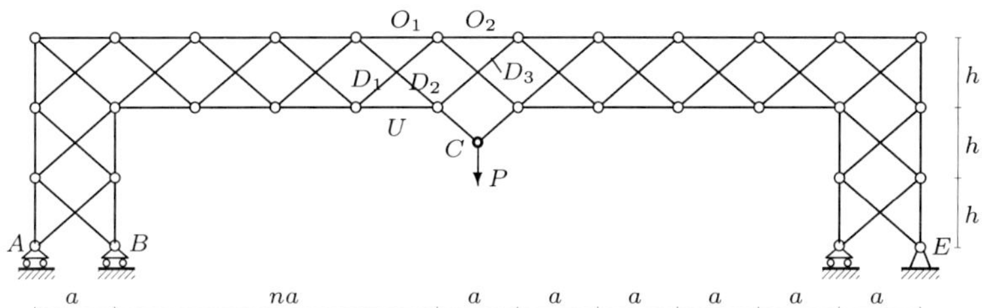

Fig. 82. Truss, $m = 2$, $n = 4$

Deflection. The coefficients in (2.32):

$C_1 = (4n^3 + 6((-1)^n - 2)n^2 + 2(7 - 6(-1)^n)n - 21(-1)^n + 30)/12,$

$C_2 = (2n^2 + 2((-1)^n - 2)n + 7 - 4(-1)^n)/4,$

$C_3 = (2(1 + 2m)n^2 + 2((-1)^n - 3 + 2(-1)^n m - 4m)n+$
$\quad +5 - 3(-1)^n - 4(-1)^n m + 6m)/4.$

Support offset. The coefficients in (2.33):

$A_1 = (2(1 + (-1)^n + (-1)^{m+n})n^2+$
$\quad +2(-2(-1)^{m+n} + (-1)^m - (-1)^n)n-$
$\quad -5(-1)^n - 8(-1)^{m+n} - 3 + 6(-1)^m)/4,$

$A_2 = (2((-1)^n + 2 + (-1)^{m+n})n+$
$\quad +5(-1)^m - 1 - 3(-1)^n - 5(-1)^{m+n})/4,$

$A_3 = (2(2m^2 + (2(-1)^{m+n} + 2(-1)^n + 4)m + (-1)^n+$
$\quad +1 - (-1)^m + (-1)^{m+n})n + 2((-1)^n - 2)m^2 + 2((-1)^m - 3-$
$\quad -2(-1)^{m+n})m - 4(-1)^{m+n} - 1 - 2(-1)^n + 3(-1)^m)/4.$

Forces:

$O_1 = -Pa/(2h)$, $O_2 = -Pa((-1)^n - 2)/(2h)$,

$D_1 = Pc((-1)^n - 1)/(4h)$,

$D_2 = Pc(1 + (-1)^n)/(4h)$,

$D_3 = Pc(1 - (-1)^n)/(4h)$,

$U = Pa(1 - (-1)^n)/(2h)$.

Supports reactions:

$Y_A = Y_E = P(3 - 2n - (-1)^n)/4$,

$Y_B = P(2n - 1 + (-1)^n)/4$, $X_E = 0$.

2.19.2. Load on the upper belt

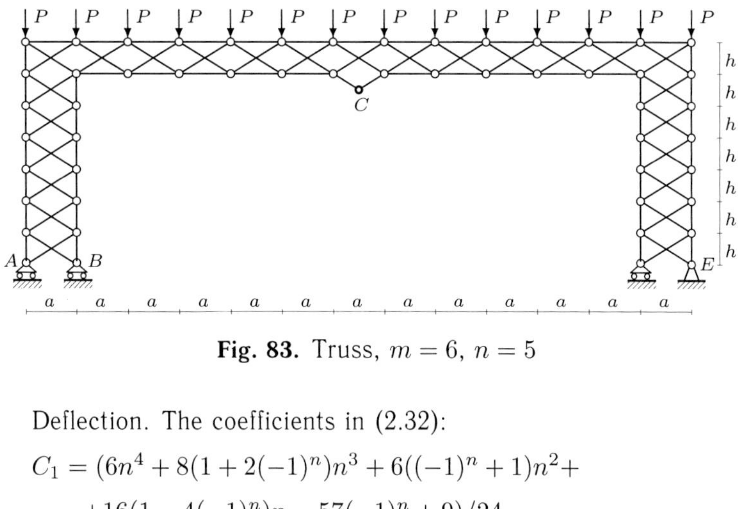

Fig. 83. Truss, $m = 6$, $n = 5$

Deflection. The coefficients in (2.32):

$$C_1 = (6n^4 + 8(1 + 2(-1)^n)n^3 + 6((-1)^n + 1)n^2 +$$
$$+16(1 - 4(-1)^n)n - 57(-1)^n + 9)/24,$$

$$C_2 = (4n^3 + 2(2 + 3(-1)^n)n^2 + 8n + 3 - 11(-1)^n)/8,$$

$$C_3 = (2(1 + 2m)n^3 + (6(-1)^n m + 4m - 1 + 3(-1)^n)n^2 +$$
$$+((-1)^n + 6(-1)^n m - 2m - 5)n + (5m + 6)(1 - (-1)^n))/4.$$

Support offset. The coefficients in (2.33):

$$A_1 = (4((-1)^{m+n} + (-1)^n + 1)n^3 +$$
$$+6(6 + 3(-1)^m + 4(-1)^n + 2(-1)^{m+n})n^2 +$$
$$+4(11(-1)^n + 20 + 2(-1)^{m+n} + 12(-1)^m)n +$$
$$+27(-1)^m + 24(-1)^n + 48 - 3(-1)^{m+n})/12,$$

$$A_2 = (((-1)^m + 3 + (-1)^{m+n} + (-1)^n)n^2 + 2((-1)^n + 2(-1)^m +$$
$$+3)n + 4 - (-1)^{m+n} + 3(-1)^m + 2(-1)^n)/2,$$

$$A_3 = ((2n^2 + 2n((-1)^n + 2) + 3(-1)^n + 1)m^2 +$$
$$+(2(2+(-1)^n+(-1)^{m+n})n^2+2((-1)^m+4(-1)^n+2(-1)^{m+n}+$$
$$+5)n + 3(-1)^m + (-1)^{m+n} + 5 + 7(-1)^n)m +$$
$$+(1+(-1)^{m+n}+(-1)^n-(-1)^m)n^2+(3+2(-1)^n-(-1)^m)n+$$
$$+2 + (-1)^m - 3(-1)^{m+n})/2.$$

Forces:

$$O_1 = Pa/h,$$
$$O_2 = Pa(1 + (3 + 2n)(-1)^n)/(2h),$$
$$D_1 = Pc((3 + 2n)(-1)^n - 3)/(4h),$$
$$D_2 = Pc(1 + (3 + 2n)(-1)^n)/(4h),$$
$$D_3 = -Pc(1 + (3 + 2n)(-1)^n)/(4h),$$
$$U = -Pa(1 + (3 + 2n)(-1)^n)/(2h).$$

Supports reactions:

$$Y_A = Y_E = -P(2n^2 + 2(1 + (-1)^n)n - 3 + 3(-1)^n)/4,$$
$$Y_B = P(2n^2 + (2(-1)^n + 6)n + 3(-1)^n + 5)/4,$$
$$X_E = 0.$$

Chapter 3
Composite Frames and Arches

Truss 3.1

A truss with a height of $(m+1)h$, $h = a\sqrt{3}/2$ consists of $8n+8m+10$ rods of length a (Fig. 84).

3.1.1. Concentrated force in the mid-span

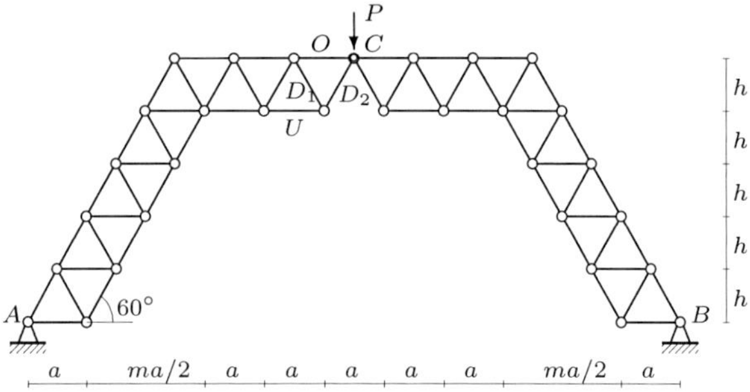

Fig. 84. Truss, $m = 4$, $n = 2$

The deflection (vertical displacement of the middle node C) has the form

$$\Delta = 2aP((2n^2+2+n)m^3 + (2n^3+9n^2+7n+9)m^2+$$
$$+(n^3+7n^2+16n+19)m+$$
$$+(2n+3)(n^2+3n+5))/(9(m+1)^2EF).$$

Forces:

$$O = -P\sqrt{3}\,(n+1)/(3(m+1)),\ D_1 = P\sqrt{3}/3,$$
$$D_2 = U = -P\sqrt{3}/3.$$

Supports reactions:

$$X_A = \sqrt{3}\,P(2n+m+3)/(6(m+1)),\ Y_A = Y_B = P/2.$$

3.1.2. Load on the upper belt

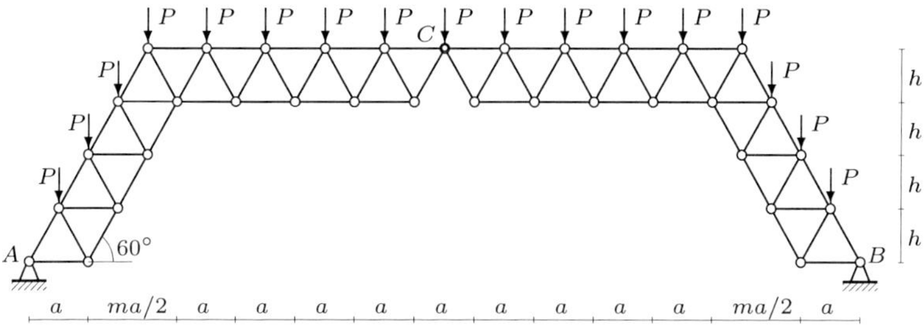

Fig. 85. Truss, $m = 3$, $n = 4$

The deflection (vertical displacement of the middle node C in the upper zone) has the form

$$\Delta = a(2(3m^2+m+4)n^4 + 2(4m^3+17m^2+13m+24)n^3+$$
$$(3m^3+24m^2+79m+130)n^2-$$
$$-(m^5+9m^4+16m^3-35m^2-169m-174)n+$$
$$+2(m+3)(4m^3+14m^2+23m+16))/(18(m+1)^2EF).$$

Forces:

$$O = \ P\sqrt{3}(2n^2+2(3+m)n+m^2+4+3m)/(6(m+1)),$$
$$D_1 = P\sqrt{3}/3,\ D_2 = U = -P\sqrt{3}/3.$$

Supports reactions:

$$X_A = P\sqrt{3}\,(2n^2+2(3+m)n+5+4m+m^2)/(6(m+1)),$$
$$Y_A = Y_B = P(2m+2n+3)/2.$$

Truss 3.2

A truss with a height of $(n+m)h$ (Fig. 86) containing $2n+2m$ panels, consists of $8n+8m+2$ rods. The case $d = a$, $b = a/2$ is considered. The deflection (vertical displacement of the middle node C) has the form

$$\Delta = P(C_1a^3 + C_2c^3 + C_3g^3 + C_4h^3)/(h^2EF(m+n)^2), \qquad (3.1)$$

where $c = \sqrt{a^2 + 4h^2}$, $g = \sqrt{a^2 + h^2}$.

The total length of the rods in the truss is

$$L_{sum} = 2(m+n+1)a + 2hn + 3cm + 4gn.$$

3.2.1. Concentrated force in the mid-span

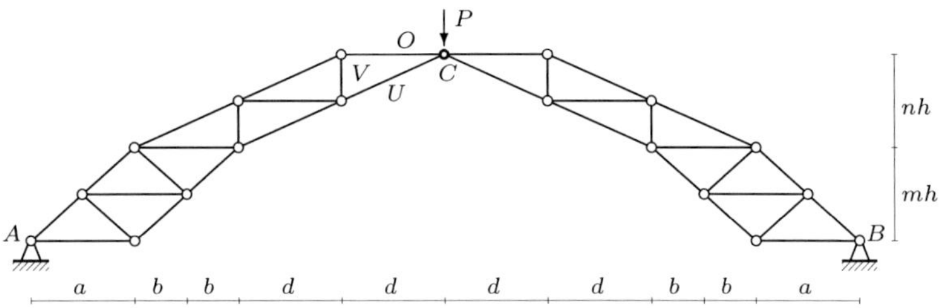

Fig. 86. Truss, $n = m = 2$

Deflection. The coefficients in (3.1):

$C_1 = (m(m+4)n^2 + 4(4+2m+m^2)n + 16m + 16)/32,$

$C_2 = m((m^2 - 3m + 11)n^2 + (m+7)(m+2)n + 8 + 6m + 4m^2)/96,$

$C_3 = n((2n^2 + 6n + 7)m^2 - 2(n+1)(n-4)m + 4(n+1)(1+2n))/24,$

$C_4 = n(m-2)^2/8.$

Forces:

$O = Pa(m-2)/(4h(m+n)),$

$U = -Pg/(2h)$, $V = P(m-2)/(4(m+n)).$

Supports reactions:

$Y_A = Y_B = P/2$, $X_A = Pa(m+2+2n)/(4h(m+n)).$

3.2.2. Load on the upper belt

Deflection. The coefficients in (3.1):

$$C_1 = -(n(n+2)m^3 + 2(3n^2+5n-4)m^2 + \\ +4(n^3+n^2-7n-6)m - 16(n+1)^2)/32,$$

$$C_2 = -m(nm^4 + (5n^2+4n-16)m^3 - (65n+40)m^2 + \\ +(12n^3-77n^2-100n-32)m - 4(n+1)(15n^2+12n+8))/384,$$

$$C_3 = -(n(n+2)(2n-1)m^3 + n(n+1)(n^2-7)m^2 + \\ +n(n+1)(n^3+5n^2-5n-10)m - 4n(2n+1)(n+1)^2)/24,$$

$$C_4 = -n(m^3-4m^2+6m-4)/8.$$

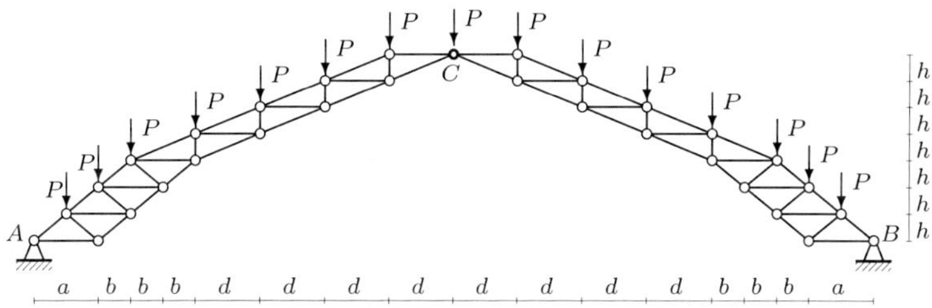

Fig. 87. Truss, $m = 3$, $n = 4$

Forces:

$$O = -Pa(m^2+2+2nm+2n+2n^2)/(4h(m+n)),$$

$$U = -Pg/(2h),\ V = P(m^2+2(n-2)m+2+2n^2-2n)/(4(n+m)).$$

Supports reactions:

$$Y_A = Y_B = P(2n+2m+1)/2,$$

$$X_A = P(am^2+2a(1+n)m+2(1+n)^2a)/(4h(m+n)).$$

Truss 3.3

A truss with a height of $2mh$ (Fig. 88) containing $2n+2m$ panels, consists of $8(n+m)-6$ rods. The total length of the rods in the truss is

$$L_{sum} = 4(2n-1)a + 4(2m-1)h + 2(2m+2n-1)c,$$

where $c = \sqrt{a^2+h^2}$.

The deflection (vertical displacement of the middle node C) has the form

$$\Delta = P(C_1a^3 + C_2c^3 + C_3h^3)/(h^2m^2EF). \tag{3.2}$$

3.3.1. Concentrated force in the mid-span

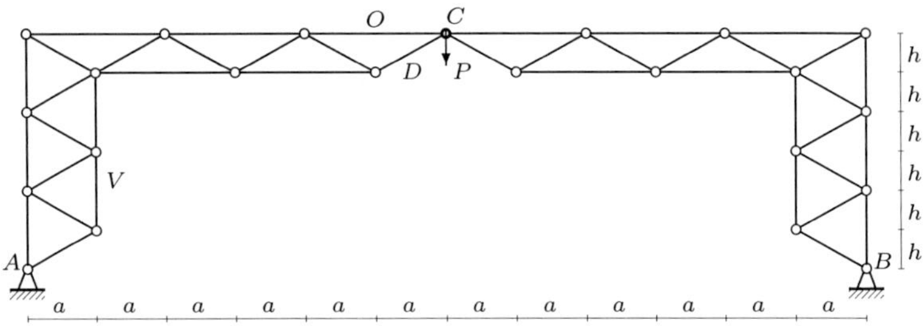

Fig. 88. Truss, $m = 3$, $n = 3$

Deflection. The coefficients in (3.2):

$C_1 = n(m^2 - 6m^2n + (8m^2 - 6m + 3)n^2)/3,$

$C_2 = mn(1 - m + (2m - 1)n),$

$C_3 = m(m^2 - 2m^2n + (8m^2 - 6m + 1)n^2/3).$

Forces $m > 1$:

$O = Pa(m - n)/(2hm)$, $D = -Pc/(2h)$, $V = -Pn/m$.

Supports reactions: $Y_A = Y_B = P/2$, $X_A = X_B = Pan/(2hm)$.

3.3.2. Load on the upper belt

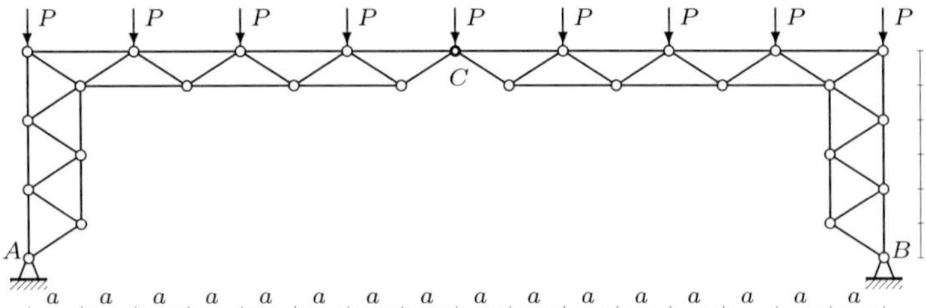

Fig. 89. Truss, $m = 3$, $n = 4$.

Deflection. The coefficients in (3.2):

$$C_1 = (3n^2(2n^2 - 2n + 1)m^2 - n^2(1 + 5n^2)m + 3n^4)/3,$$

$$C_2 = m(2n(2n^2 - 2n + 1)m - n(2n - 1)(n - 1))/2,$$

$$C_3 = m((8n^3 - 9n^2 + 3n + 3)m^2 - 6n^3m + n^3)/3.$$

Forces $m > 1$:

$$O = Pa(m - n^2)/(2hm),\ D = -Pc/(2h),\ V = -Pn^2/m.$$

Supports reactions:

$$Y_A = Y_B = P(2n + 1)/2,\ X_A = X_B = Pan^2/(2hm).$$

Truss 3.4

A truss with a height of $(m + 1)h$ (Fig. 90) containing $2n + 2m$ panels, consists of $8(n + m) + 10$ rods. The total length of the rods in the truss is

$$L_{sum} = 2(2n + m + 1)a + (2m + 4n + 7)c + 2(2m - 1)h,$$

where $c = \sqrt{a^2 + h^2}$.

The deflection (vertical displacement of the middle node C) has the form $(m > 1)$:

$$\Delta = P(C_1a^3 + C_2c^3 + C_3h^3)/(2(m + 1)^2h^2EF). \qquad (3.3)$$

3.4.1. Concentrated force in the mid-span

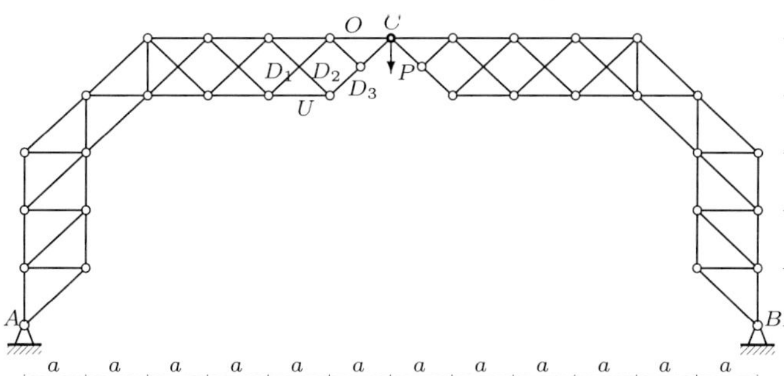

Fig. 90. Truss, $m = 4$, $n = 3$

Deflection. The coefficients in (3.3):

$$C_1 = (2(n+2)(2n^2+5n+9)m^2 + (2n^3 - 12n^2 - 3n(-1)^n - \\ -65n - 9(-1)^n - 81)m - 3n(-1)^n + \\ +4n^3 + 24n^2 + 59n - 9(-1)^n + 45)/6,$$

$$C_2 = (n+2)(3n+5)m^2 - (n^2+10n+13)m + 13 + 3n,$$

$$C_3 = (2(2n^2+9n+12)m^3 - (6n^2+54n+6n(-1)^n + \\ +81 + 9(-1)^n)m^2 + 2(7n^2+30n-3n(-1)^n+45)m - \\ -21 - 12n + 9(-1)^n)/6.$$

Forces:

$$O = -Pa(n-m+2)/(2h(m+1)),$$

$$D_1 = 0,\ D_2 = Pc/(2h),\ D_3 = -Pc/(2h),\ U = -Pa/h.$$

Supports reactions:

$$Y_A = Y_B = P/2,\ X_A = Pa(n+3)/(2(m+1)h).$$

3.4.2. Load on the upper belt

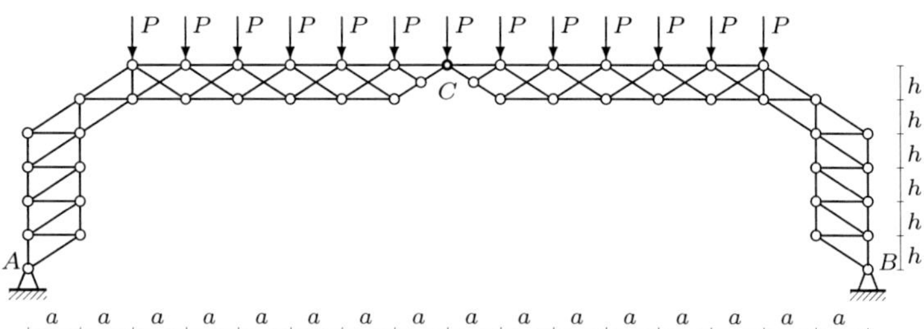

Fig. 91. Truss, $m = n = 5$

Deflection. The coefficients in (3.3):

$$C_1 = (2(3m^2+m+4)n^4 + 36(m^2-m+2)n^3 + \\ +(108m^2 - 251m + 265 - 3(-1)^n m - 3(-1)^n)n^2 + \\ +(192m^2 - 18(-1)^n m - 546m + 414 - 18(-1)^n)n + \\ +3((-1)^n + 47)m^2 - 3(5(-1)^n + 127)m - 18(-1)^n + 234)/12,$$

$$C_2 = ((12n^3 + 62n^2 + 112n + 71 + (-1)^n)m^2 + (2(-1)^n - 126-$$
$$-176n - 68n^2 - 4n^3)m + 14n^2 + 112n + (-1)^n + 139)/4,$$
$$C_3 = ((4n^3 + 27n^2 + 67n + 54)m^3 - 3(2n^3 + ((-1)^n + 27)n^2+$$
$$+(3(-1)^n+77)n+63+3(-1)^n)m^2+(14n^3+3(30-(-1)^n)n^2+$$
$$+239n + 3(-1)^n + 201)m - 18n^2+$$
$$+9((-1)^n - 7)n - 60 + 12(-1)^n)/6.$$

Forces:

$O = -Pa(n^2 + 6n - m + 6)/(2h(m + 1)),$

$D_1 = -Pc/h,\ D_2 = Pc/(2h),\ D_3 = -Pc/(2h),\ U = -Pa/h.$

Supports reactions:

$X_A = X_B = Pa(n^2 + 6n + 7)/(2h(m + 1)),$

$Y_A = Y_B = P(2n + 3)/2.$

Truss 3.5

A truss with a height of $(m + 1)h$ (Fig. 92) containing $2n + 2m$ panels, consists of $8(n + m) + 12$ rods.

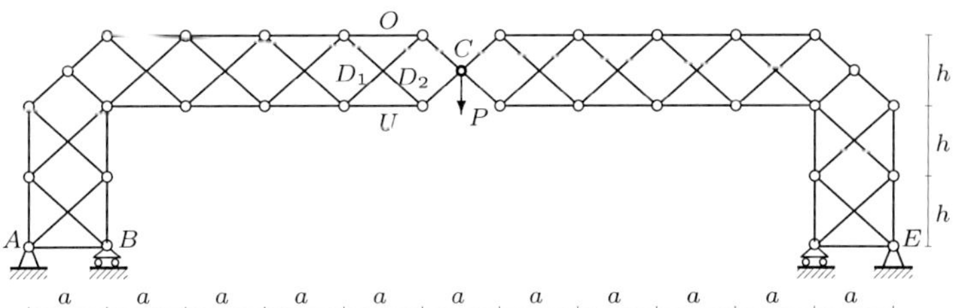

Fig. 92. Truss, $m = 2$, $n = 4$

The total length of the rods in the truss is

$$L_{sum} = 2(1 + 2n)a + (4m + 4n + 5)c + 4hm,$$

where $c = \sqrt{a^2 + h^2}$.

The deflection (vertical displacement of the central node C) has the form

$$\Delta = P(C_1 a^3 + C_2 c^3 + C_3 h^3)/(h^2 EF), \tag{3.4}$$

where $c = \sqrt{a^2 + h^2}$. The truss is unchangeable when $n = 2, 4, 6, ...$ or when $n = 2k$, $k = 1, 2, 3, ...$. With $n = 1, 3, 5, ...$ the truss is kinematically variable (Fig. 93).

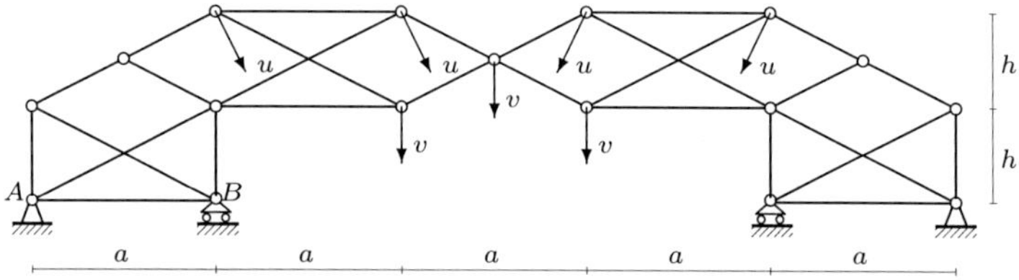

Fig. 93. Diagram of possible velocity of nodes for an instantly changeable truss, $m = 1$, $n = 1$, $v/(2a) = u/c$

3.5.1. Concentrated force in the mid-span

Deflection. The coefficients in (3.4):

$$C_1 = (80k^3 + 6(17 + 4(-1)^m)k^2 + 2(17 + 9(-1)^m)k +$$
$$+3(-1)^m + 3)/12,$$

$$C_2 = (8k^3 + 12k^2 + 6k + 2(10k^2 + 6k + 1)m + 1)/4,$$

$$C_3 = (4(2k + 1)^2 m^3 - 6(2k + 1)^2 m^2 +$$
$$+8(4k + 7k^2 + 1)m - 3 + 12(-1)^m k + 3(-1)^m - 12k)/12.$$

Forces:

$O = -Pak/h$, $D_1 = Pck/(2h)$,

$D_2 = Pc(k + 1)/(2h)$, $U = -Pa(k + 1)/h$.

Supports reactions:

$X_A = Pa(1 + 2k)/(2h)$,

$Y_A = P((2m - 1)k + m)/2$,

$Y_B = ((1 - 2m)k + 1 - m)/2$.

3.5.2. Load on the upper belt

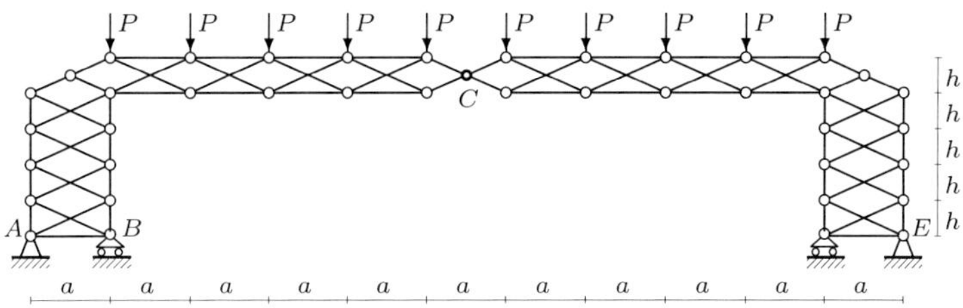

Fig. 94. Truss, $m = n = 4$

Deflection. The coefficients in (3.4):

$$C_1 = (48k^4 + 4(4(-1)^m + 21)k^3 + 2(9(-1)^m + 23)k^2 +$$
$$+2(5 + 4(-1)^m)k + 1 + (-1)^m)/4,$$
$$C_2 = (16k^4 + 8(4 + 5m)k^3 + 2(11 + 18m)k^2 + (1 + 2m)(6k + 1))/4,$$
$$C_3 = (16m(2m^2 - 3m + 7)k^3 + 12(4m^3 - 6m^2 + 8m + 3(-1)^m -$$
$$-3)k^2 + 4(8m^3 - 9m^2 + 7m + 6(-1)^m - 6)k + 8m^3 - 6m^2 +$$
$$+4m + 3(-1)^m - 3)/12.$$

Forces:

$O = -2Pa(k + 1)k/h$, $D_1 = Pc(2k^2 + 2k - 1)/(2h)$,

$D_2 = Pc(1 + 2k + 2k^2)/(2h)$, $U = -Pa(1 + 2k + 2k^2)/h$.

Supports reactions:

$X_A = Pa(1+2k+2k^2)/h$, $Y_A = P(2(2m-1)k^2+4km+2m+1)/2$,

$Y_B = P(2(1 - 2m)k^2 + 4(1 - m)k - 2m + 1)/2$.

Truss 3.6

A truss with a height of $(m + 2)h$ consists of $8m + 8n + 32$ rods (Fig. 95). The total length of the rods in the truss is

$$L_{sum} = 4(n + 2)a + 2(3m + 9 + 2n)c + 4(m + 1)h,$$

where $c = \sqrt{a^2 + h^2}$. Case $m = 3$ is considered.

The deflection (vertical displacement of the middle node C in the upper zone) has the form [21]

$$\Delta = P(C_1a^3 + C_2c^3 + C_3h^3)/(h^2EF). \qquad (3.5)$$

3.6.1. Concentrated force in the mid-span

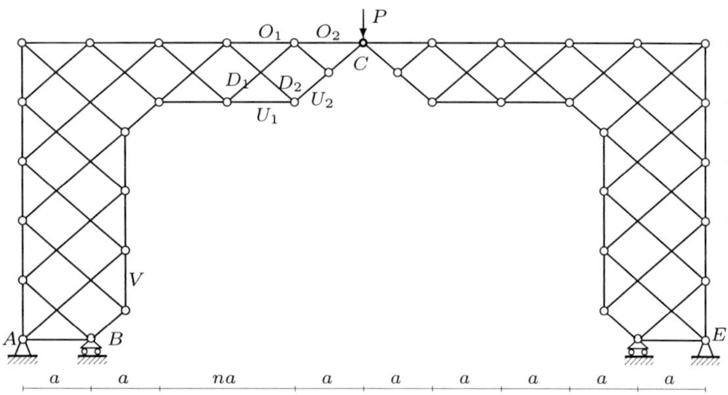

Fig. 95. Truss, $m = 3$, $n = 2$

Deflection. The coefficients in (3.5):

$C_1 = (8n^3 + 30n^2 + (43 - 12(-1)^n)n - 15(-1)^n + 27)/24,$

$C_2 = (60n^2 + 6(-2(-1)^n + 15)n + 129 - 63(-1)^n)/48,$

$C_3 = (60n^2 + 12(6 - (-1)^n)n + 75 - 24(-1)^n)/24.$

Forces:

$O_1 = O_2 = Pa/(4h),$

$D_1 = 0,\ D_2 = Pc/(2h),$

$U_1 = -Pa/h,\ U_2 = -Pc/(2h),$

$V = -P((-1)^n + 1 + 2n)/4.$

Supports reactions:

$X_A = Pa/(4h),\ Y_A = Y_E = P(1 - 2n)/4,$

$Y_B = P(1 + 2n)/4.$

3.6.2. Load on the upper belt

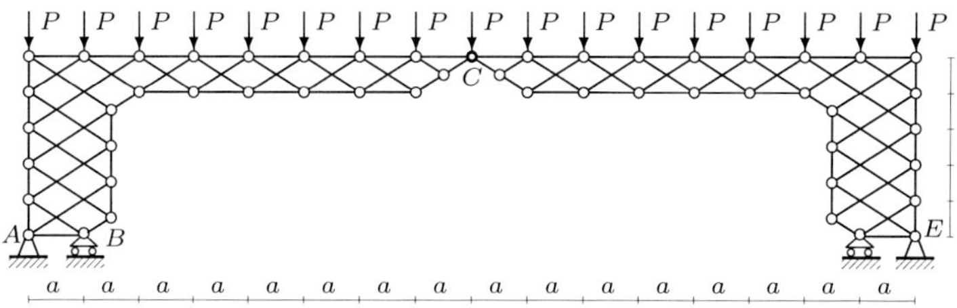

Fig. 96. Truss, $m = 3$, $n = 5$

Deflection. The coefficients in (3.5):

$$C_1 = (12n^4 + 56n^3 + 6(15 - 2(-1)^n)n^2 + 2(32 - 15(-1)^n)n+$$
$$+33 - 9(-1)^n)/48,$$
$$C_2 = (40n^3 + 2(37 - 2(-1)^n)n^2 - (22 + 42(-1)^n)n+$$
$$+67 - 55(-1)^n)/32,$$
$$C_3 = (20n^3 + 2(18 - (-1)^n)n^2 - 2(23 + 4(-1)^n)n + 16 + (-1)^n)/8.$$

Forces:

$$O_1 = -Pa(2n - 1)/(4h),\ O_2 = -Pa(3 + 2n)/(4h),$$
$$D_1 = -Pc/h,\ D_2 = Pc/(2h),\ U_1 = -Pa/h,$$
$$U_2 = -Pc/(2h),\ V = -P(n^2 + n - 3)/2.$$

Supports reactions:

$$X_A = Pa(5 + 2n)/(4h),\ Y_A = Y_E = P(21 + 2n - 2n^2)/4,$$
$$Y_B = P(2n^2 + 2n - 7)/4.$$

Truss 3.7

A truss with a height of $(m + 1.5)h$ consists of $8(m + n + 3)$ rods (Fig. 97). The total length of the rods in the truss is

$$L_{sum} = 2(2n + 3)a + 2(3m + 2n + 6)c + 2(2m + 1)h,$$

where $c = \sqrt{a^2 + h^2}$.

The deflection (vertical displacement of the middle node C in the upper zone) has the form

$$\Delta = P(C_1a^3 + C_2c^3 + C_3h^3)/(h^2EF). \quad (3.6)$$

Case $m = 3$ is considered.

3.7.1. Concentrated force in the mid-span

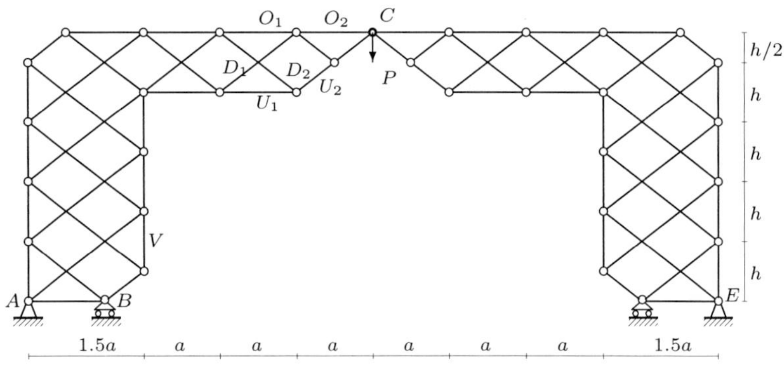

Fig. 97. Truss, $m = 3$, $n = 2$

Deflection. The coefficients in (3.6):

$C_1 = (12n^3 + 48n^2 + (74 - 18(-1)^n)n - 27(-1)^n + 45)/36$,

$C_2 = (54n^2 + (136 - 12(-1)^n)n + 151 - 75(-1)^n)/48$,

$C_3 = (30n^2 + 6(9 - (-1)^n)n + 47 - 15(-1)^n)/12$.

Forces:

$O_1 = O_2 = Pa/(3h)$, $D_1 = 0$, $D_2 = Pc/(2h)$,

$U_1 = -Pa/h$, $U_2 = -Pc/(2h)$,

$V = -P(5 + 6n + 3(-1)^n)/12$.

Supports reactions:

$X_A = Pa/(6h)$, $Y_A = Y_E = -Pn/2$,

$Y_B = P(1 + n)/2$.

3.7.2. Load on the upper belt

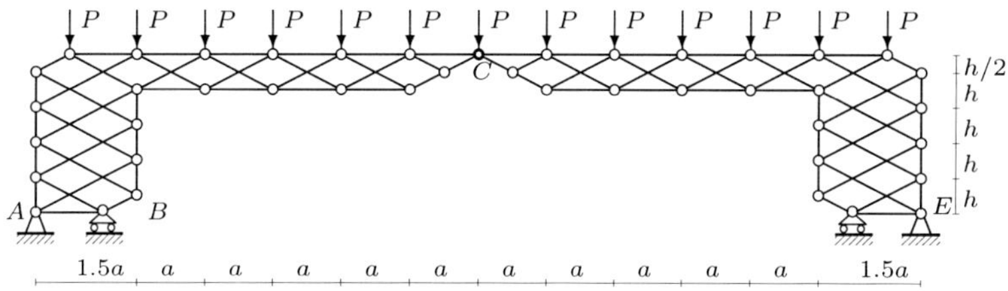

Fig. 98. Truss, $m = 3$, $n = 4$

Deflection. The coefficients in (3.6):

$$C_1 = (9n^4 + 46n^3 + (88 - 9(-1)^n)n^2 +$$
$$+3(25 - 9(-1)^n)n + 24 - 18(-1)^n)/36,$$
$$C_2 = (18n^3 + (60 - 2(-1)^n)n^2 +$$
$$+(57 - 25(-1)^n)n - 39(-1)^n + 35)/16,$$
$$C_3 = (30n^3 + 3(27 - (-1)^n)n^2 +$$
$$+(31 - 15(-1)^n)n - 12(-1)^n + 28)/12.$$

Forces:

$O_1 = -Pa(n-2)/(3h)$, $O_2 = -Pa(1+n)/(3h)$,

$D_1 = -Pc/h$, $D_2 = Pc/(2h)$,

$U_1 = -Pa/h$, $U_2 = -Pc/(2h)$, $V = -P(3n^2 + 5n - 4)/6$.

Supports reactions: $X_A = Pa(5 + 2n)/(6h)$,

$Y_A = Y_E = P(6 - n^2)/2$, $Y_B = P(n^2 + 2n - 1)/4$.

Truss 3.8

A truss with a height of $(m+1)h$ consists of $8m + 8n + 16$ rods (Fig. 99). The total length of the rods in the truss is

$$L_{sum} = 4(n+1)a + (4m + 4n + 7)c + 2(1 + 2m)h,$$

where $c = \sqrt{a^2 + h^2}$.

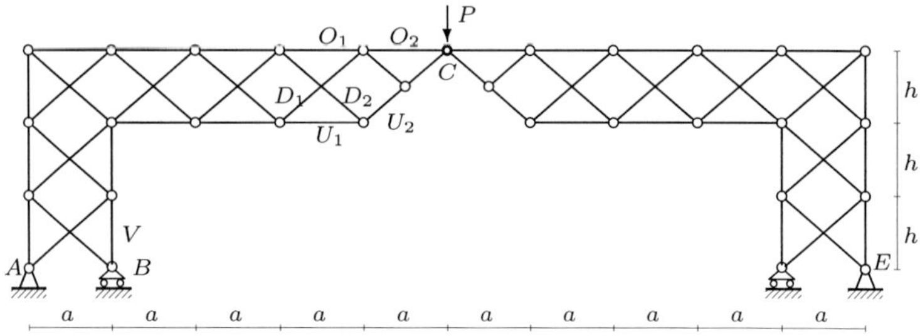

Fig. 99. Truss, $m = 2l = 2$, $n = 2k - 1 = 3$

The truss is unchangeable when $n = 2k - 1$, $k = 1, 2, 3, ..$ and $m = 2l$, $l = 1, 2, 3, ..$. Otherwise the truss is kinematically variable (Fig. 100).

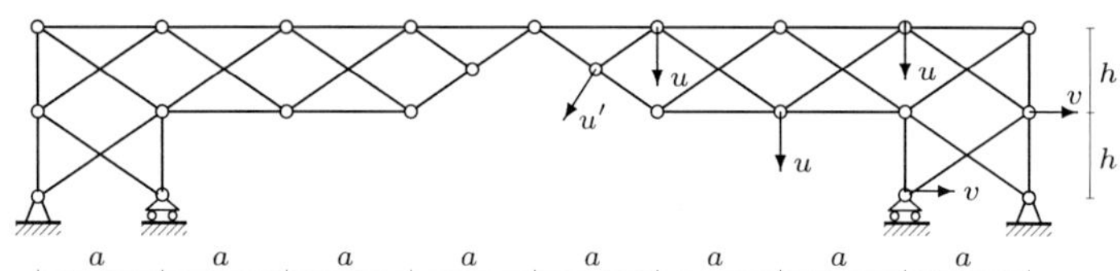

Fig. 100. Diagram of possible velocity of nodes for an instantly changeable truss, $m = 1$, $n = 2$, $v/h = u/a = 2u'/c$

The deflection (vertical displacement of the middle node C) has the form

$$\Delta = P(C_1 a^3 + C_2 c^3 + C_3 h^3)/(h^2 EF). \tag{3.7}$$

3.8.1. Concentrated force in the mid-span

Deflection. The coefficients in (3.7):

$C_1 = 4k(1 + 2k^2)/3$, $C_2 = (2l + 4k^2 + 2k + 1)/2$,

$C_3 = (12k^2(4l + 1) - 24l(2l + 1)k + 8l + 16l^3 + 12l^2)/6$.

Forces:

$O_1 = O_2 = D_1 = 0$, $D_2 = Pc/(2h)$,

$U_1 = -Pa/h$, $U_2 = -Pc/(2h)$, $V = P(l - k)$.

Supports reactions:

$X_A = Pa/(2h)$, $Y_A = P((2l+1)/2 - k)$, $Y_B = P(k-l)$.

3.8.2. Load on the upper belt

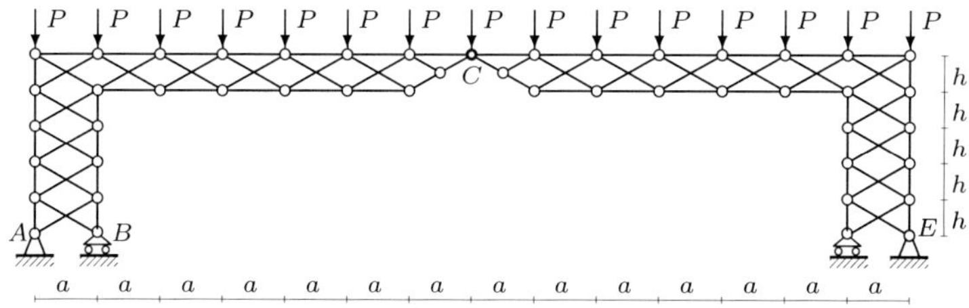

Fig. 101. Truss, $m = 2l = 4$, $n = 2k - 1 = 5$

Deflection. The coefficients in (3.7):

$C_1 = k(12k^3 + 2k^2 + 3k + 1)/3$,

$C_2 = (8k^3 + 2k^2 + (4l+2)k + 2l + 1)/2$,

$C_3 = (12(1+4l)k^3 - 12l(1+4l)k^2 + (16l^3 - 6l^2 - 10l - 6)k +$
$+2l(4l^2 + 6l + 5))/3$.

Forces:

$O_1 = -Pa(k-1)/h$, $O_2 = -Pak/h$, $D_1 = -Pc/h$, $D_2 = Pc/(2h)$,

$U_1 = -Pa/h$, $U_2 = -Pc/(2h)$, $V = -P(2k^2 - (2l-1)k - l)$.

Supports reactions:

$X_A = Pa(1+2k)/(2h)$, $Y_A = P(3 + 2l - 4k^2 + 2(1+2l)k)/2$,

$Y_B = P(4k^2 + 2(1-2l)k - 2l)/2$.

Truss 3.9

A truss with a height of $(2m+1)h/2$ consists of $8(m+n+1)$ rods (Fig. 102). The total length of the rods in the truss is

$$L_{sum} = 2(2n+1)a + (4m+2n+3)c + 4hm,$$

where $c = \sqrt{a^2 + h^2}$.

The deflection (vertical displacement of the middle node C) has the form

$$\Delta = P(C_1a^3 + C_2c^3 + C_3h^3)/(h^2EF). \tag{3.8}$$

3.9.1. Concentrated force in the mid-span

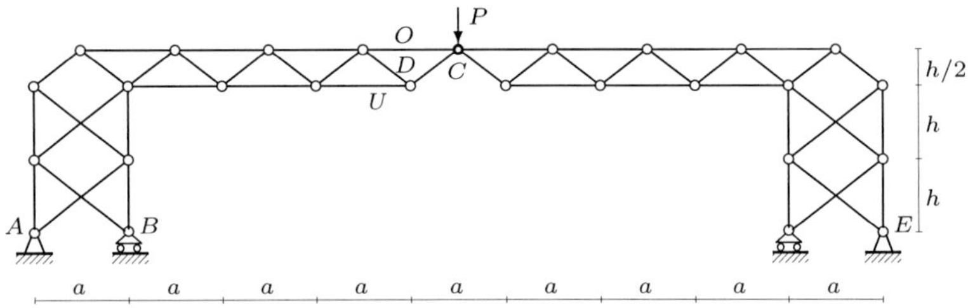

Fig. 102. Truss, $m = 2$, $n = 3$

Deflection. The coefficients in (3.8):

$C_1 = ((n+1)(2n+1)(3n(-1)^m + 5n + 3(-1)^m + 3))/3,$

$C_2 = (2(2n+1)^2m + (2n+1)(2n(-1)^m + 2n + (-1)^m + 2))/4,$

$C_3 = (4(2n+1)^2m^3 + 6(2n+1)(2((-1)^m + 1)n + 2(-1)^m + 1)m^2 +$
$+4((6(-1)^m + 8)n^2 + (9(-1)^m + 11)n + 3(-1)^m + 5)m +$
$+3(2n+1)((-1)^m - 1))/12.$

Forces:

$O = Pa(2(-1)^m n + (-1)^m + 1)/(2h),$

$D = Pc/(2h),$

$U = -Pa/h,$

$V = -P((-1)^m(1+2m)(1+2n) + 2n + 3)/4.$

Supports reactions:

$X_A = -Pa(-1)^m(2n+1)/(2h),$

$Y_A = -P(1+2n)(1+(-1)^m(1+2m))/4,$

$Y_B = P((-1)^m(1+2m)(1+2n) + 2n + 3)/4.$

3.9.2. Load on the upper belt

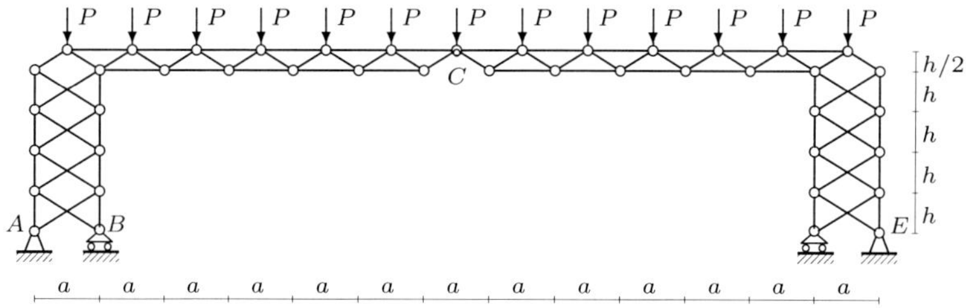

Fig. 103. Truss, $m = 4$, $n = 5$

Deflection. The coefficients in (3.8):

$$C_1 = (2(9 + 5(-1)^m)n^4 + 16(3 + 2(-1)^m)n^3 + (39 + 35(-1)^m)n^2 + \\ + (9 + 13(-1)^m)n)/6,$$

$$C_2 = (8((-1)^m + 2m + 1)n^3 + 4(4 + 3(-1)^m + 6m)n^2 + \\ + 4(2 + (-1)^m)n + 2 - 4m)/8,$$

$$C_3 = (4(2n + 1)(2n^2 + 2n - 1)m^3 + 6(4((-1)^m + 1)n^3 + \\ + 3(3(-1)^m + 2)n^2 + 5n(-1)^m - 1)m^2 + 2(4(4 + 3(-1)^m)n^3 + \\ + 3(9(-1)^m + 11)n^2 + 3(5(-1)^m + 7)n + 8)m + \\ + 3(3n^2 + 5n + 1)((-1)^m - 1))/12.$$

Forces:

$$O = Pa(1 - (-1)^m + 2(-1)^m n + 2(-1)^m n^2)/(2h),$$

$$D = Pc/(2h),$$

$$U = -Pa/h,$$

$$V = P((-1)^m(2m + 1)(1 - 2n - 2n^2) - 2n^2 - 5 - 6n)/4.$$

Supports reactions:

$$X_A = Pa(-1)^m(1 - 2n - 2n^2)/(2h),$$

$$Y_A = -P(2n^2 + 2n - 1)(2(-1)^m m + 1 + (-1)^m)/4,$$

$$Y_B = P((2m + 1)(2n^2 + 2n - 1)(-1)^m + 2n^2 + 6n + 5)/4.$$

Truss 3.10

A truss with a height of $(m+1)h$ consists of $8m+8n+22$ rods (Fig. 104). The total length of the rods in the truss is

$$L_{sum} = 2(3+2n)a + (8m+4n+11)c + 2(2m+1)h,$$

where $c = \sqrt{a^2+h^2}$.

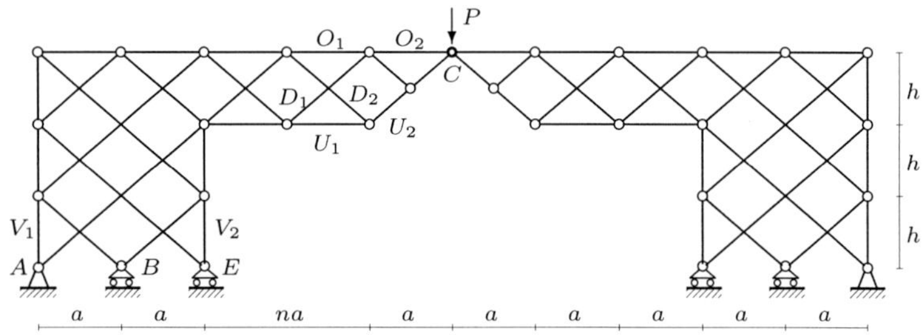

Fig. 104. Truss, $m=2$, $n=2$

The deflection (vertical displacement of the middle node C in the upper zone) has the form

$$\Delta = P(C_1a^3 + C_2c^3 + C_3h^3)/(h^2EF). \tag{3.9}$$

The truss is unchangeable when $n = 2k$, $k = 1,2,3...$ and $m = 2l$, $l = 1,2,3..$ or when $n = 2k-1$ and $m = 1,2,4,5,6,8,9,...,$ $(m = (\sqrt{3}\sin(2l\pi/3) + 3\cos(2l\pi/3) + 12l - 3)/9$, $l = 1,2,3..)$. Otherwise the truss is kinematically variable (Fig. 105).

The case $n = 2k$, $m = 2l$ is obtained.

Deflection. The coefficients in (3.9):

$C_1 = 2(k+1)(2k+1)(3(-1)^l(k+1) + 5k + 3)/3,$

$C_2 = (2k+1)(2k(1+2l) + (2k+1)(-1)^l + 2l + 2)/2,$

$C_3 = (4(2k+1)^2l^3 + 6(2k+1)(2k + 2(-1)^l(k+1) + 1)l^2+$
$+4(2(3(-1)^l + 4)k^2 + (9(-1)^l + 11)k + 5 + 3(-1)^l)l+$
$+3(2k+1)((-1)^l - 1))/6.$

3.10.1. Concentrated force in the mid-span

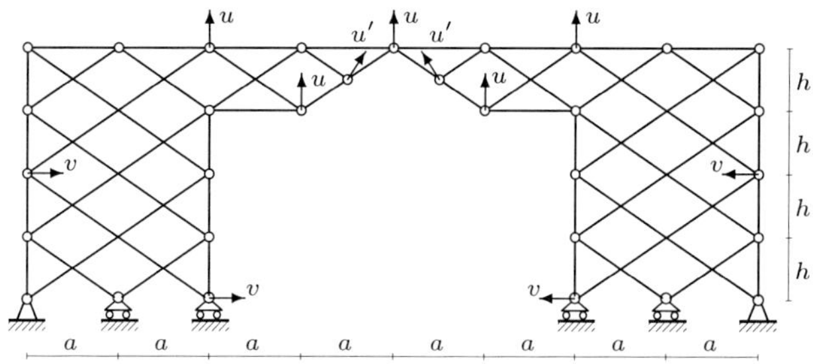

Fig. 105. Diagram of possible velocity of nodes for an instantly changeable truss, $m = 3$, $n = 1$, $v/h = u/a = 2u'/c$

Forces:

$O_1 = O_2 = Pa(1 + (-1)^l(1 + 2k))/(2h)$, $D_1 = 0$, $D_2 = Pc/(2h)$,

$U_1 = -Pa/h$, $U_2 = -D_2$, $V_1 = P(1 + 2k)(1 - (-1)^l(1 - 2l))/4$,

$V_2 = -P((-1)^l(1 + 2l)(1 + 2k) + 3 + 2k)/4$.

Supports reactions:

$X_A = -Pa(-1)^l(1 + 2k)/(2h)$,

$Y_A = -P(1 + 2k)((-1)^l(1 + 2l) + 1)/4$,

$Y_B = 0$, $Y_E = P((-1)^l(1 + 2l)(1 + 2k) + 2k + 3)/4$.

3.10.2. Load on the upper belt

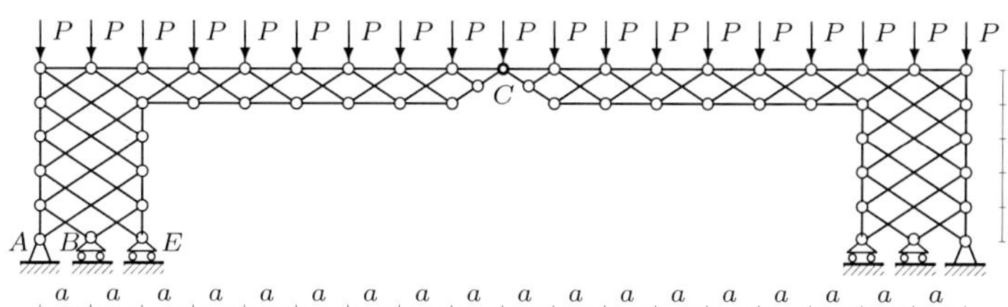

Fig. 106. Truss, $m = 6$, $n = 4$

Deflection. The coefficients in (3.9):

$$C_1 = (k+1)(36k^3 + 20(-1)^l k^3 + 60k^2 + 44(-1)^l k^2 +$$
$$+24k + 29(-1)^l k + 3 + 3(-1)^l)/3,$$
$$C_2 = ((8(-1)^l + 16l + 8)k^3 + (12(-1)^l + 24l + 14)k^2 +$$
$$+(4(-1)^l + 4l + 6)k - 2l + 1)/2,$$
$$C_3 = (16l(2l^2 + 3(1 + (-1)^l)l + 3(-1)^l + 4)k^3 +$$
$$+6(8l^3 + 6(2 + 3(-1)^l)l^2 + 2(11 + 9(-1)^l)l - 3 + 3(-1)^l)k^2 +$$
$$+(8l^3 + 6(9(-1)^l + 2)l^2 + (54(-1)^l + 82)l - 33(1 - (-1)^l))k -$$
$$-6l^2 - 4l^3 + 28l - 9 + 9(-1)^l)/6.$$

Forces:

$$O_1 = Pa(3 + (-1)^l(4k^2 + 4k - 1))/(2h),$$
$$O_2 = Pa((-1)^l(4k^2 + 4k - 1) + 1)/(2h),$$
$$D_1 = -Pc/h,\ D_2 = Pc/(2h),$$
$$U_1 = -Pa/h,\ U_2 = -D_2,$$
$$V_1 = P(8(1 + (-1)^l(2l - 1))k^2 + 8(1 + 2(-1)^l l)k +$$
$$+2(5 - 2l)(-1)^l - 10)/8,$$
$$V_2 = -P(4((-1)^l + 1 + 2(-1)^l l)k^2 + 4(2(-1)^l l + 3)k -$$
$$-5(-1)^l + 9 - 2(-1)^l l)/4.$$

Supports reactions:

$$X_A = -Pa(-1)^l(-1 + 4k + 4k^2)/(2h),$$
$$Y_A = -P(4(1 + (-1)^l(1 + 2l))k^2 + 4(1 + 2(-1)^l(1 + l))k +$$
$$+(3 - 2l)(-1)^l - 5)/4,$$
$$Y_B = 2P(-1)^l(k+1),$$
$$Y_E = P(4((-1)^l(1 + 2l) + 1)k^2 + 4(2(-1)^l l + 3)k -$$
$$-(5 + 2l)(-1)^l + 9)/4.$$

Truss 3.11

A truss with a height of $(2m+k)h$ consists of $8(m+n+k)-2$ rods (Fig. 107).

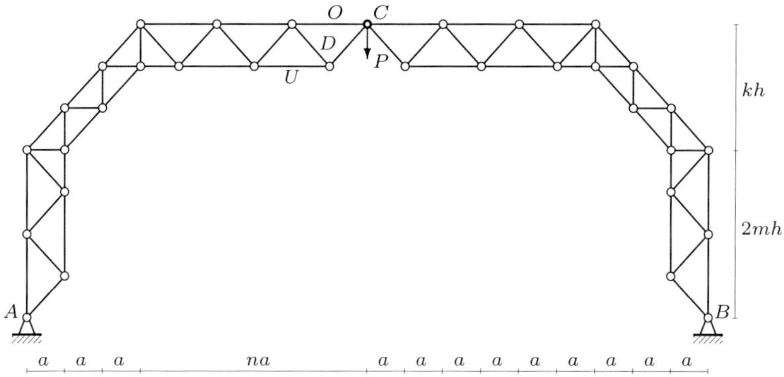

Fig. 107. Truss, $m=2$, $n=k=3$

The total length of the rods in the truss is

$$L_{sum}=2(k+4n-1)a+2(2n+2m+2k-1)c+2(4m+k-1)h,$$

where $c=\sqrt{a^2+h^2}$.

The deflection (vertical displacement of the middle node C in the upper zone) has the form

The deflection (vertical displacement of the middle node C in the lower zone) has the form

$$\Delta=P(C_1a^3+C_2c^3+C_3h^3)/((k+2m)^2h^2EF). \qquad (3.10)$$

3.11.1. Concentrated force in the mid-span

Deflection. The coefficients in (3.10):

$$C_1=(2(3k^2+3(4n-1)k+2n(4n+5)(2n-1))m^2+$$
$$+4n((8n^2-3n+1)k-6n(n-1))m+2n((4n^2-3n+2)k^2-$$
$$-3n(2n-3)k+6n^2))/3,$$

$$C_2=((48(2k-1)n^2+24(2k^2-2k+1)n+4k(2k-1)(k-1))m^2+$$
$$+(24(2k^2-2k+1)n^2+4k(2k^2-9k+13)n-6k^2(k-2))m+$$

$+k(4(2k-1)(k-1)n^2-6k(k-2)n+3k^2))/6,$

$C_3=(4(2k^2+(8n-3)k+8n^2+3-6n)m^3-$
$-6(k^2+(2n-3)k-4n(n+1))m^2+4(k^2+n(1+6n)k-5n^2)m+$
$+6kn^2(k-1))/3.$

Case $m=k=n$:

$C_1=C_3=12n^3(2n^2+1)$, $C_2=9n^3(8n^2-8n+5)/2$.

Forces:

$O=-Pa(n-m)/(h(k+2m))$, $D=Pc/(2h)$, $U=-Pa/h$.

Supports reactions:

$X_A=Pa(k+2n)/(2h(k+2m))$, $Y_A=Y_B=P/2$,

3.11.2. Load on the upper belt

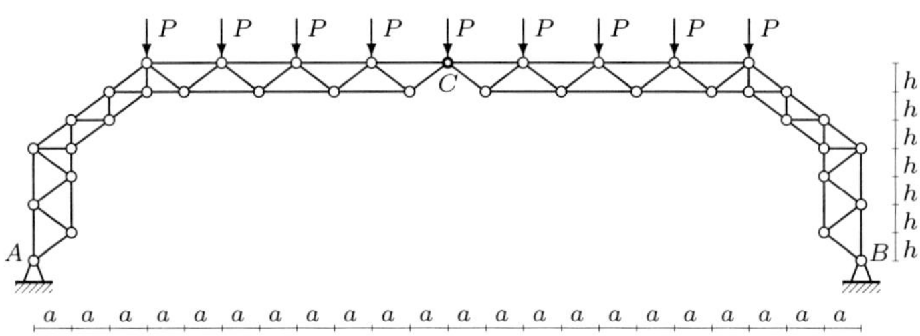

Fig. 108. Truss, $m=2$, $k=3$, $n=4$

Deflection. The coefficients in (3.10):

$C_1=(6(4n^4+4n^3+2(3k-2)n^2+2(k-1)(k+1)n+k(k-1))m^2+$
$+2n(2(6k-5)n^3+4(3-2k)n^2+(3k-2)n-k)m+$
$+2n((6-5k+3k^2)n^3-4k(k-3)n^2+k(3k+2)n+k^2))/3,$

$C_2=(4(12(2k-1)n^3+6(3k^2-3k+1)n^2+4k(k^2-1)n+$
$+k(2k-1)(k-1))m^2+2(12(2k^2-2k+1)n^3+$
$+3k(2k^2-9k+11)n^2-k(k+1)(4k-7)n-3k^2(k-2))m+$
$+k(4(2k-1)(k-1)n^3-3k(3k-5)n^2+3k(1+k)n+3k^2))/6,$

$$C_3 = (4(8n^3 + 3(4k - 3)n^2 + (4k^2 - 2k + 3)n + 2k^2 - 3k + 3)m^3 + \\ +6(4n^3 + 3(2 - k)n^2 - (2k^2 - 5k - 2)n - k(k - 3))m^2 + \\ +2(2(6k - 5)n^3 + 3n^2k + k(1 + 4k)n + 2k^2)m + 6kn^3(k - 1))/3.$$

Case $m = k = n$:

$C_1 = 6n^3(3n^3 + n^2 + 3n - 1)$, $C_2 = 3n^3(28n^3 - 27n^2 + 12n + 4)/2$,

$C_3 = 4n^3(8n^3 - n^2 + 4n + 4)$.

Forces:

$O = -Pa(n^2 + kn - m)/(h(k + 2m))$, $D = Pc/(2h)$, $U = -Pa/h$.

Supports reactions:

$X_A = Pa(k + 2n^2 + 2kn)/(2h(2m + k))$, $Y_A = Y_B = (2n + 1)P/2$.

Truss 3.12

A truss with a height of $(m + 1)h$ consists of $8m + 8n + 16$ rods (Fig. 109). The total length of the rods in the truss is

$$L_{sum} = 4(n + 1)a + (4m + 2n + 7)c + 2(2m + 1)h,$$

where $c = \sqrt{a^2 + h^2}$.

The deflection (vertical displacement of the middle node C in the upper zone) has the form

$$\Delta = P(C_1a^3 + C_2c^3 + C_3h^3)/(h^2EF). \qquad (3.11)$$

3.12.1. Concentrated force in the mid-span

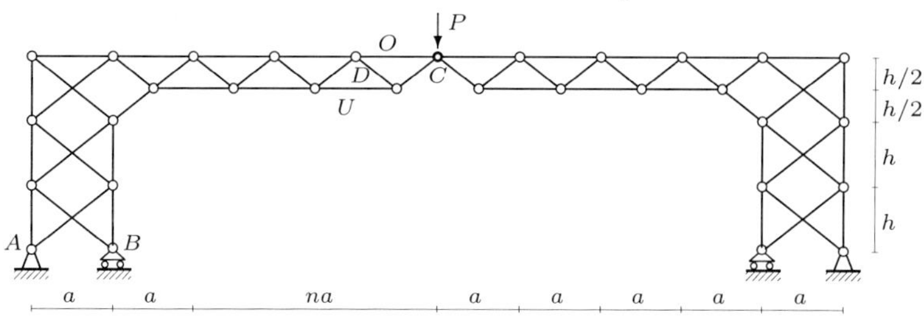

Fig. 109. Truss, $m = 2$, $n = 3$

The truss is unchangeable when $m = 2k$, $k = 1, 2, 3, \ldots$. Otherwise the truss is kinematically variable (Fig. 110).

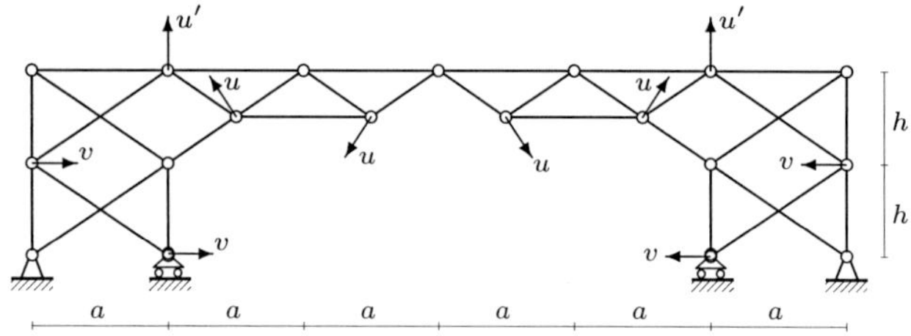

Fig. 110. Diagram of possible velocity of nodes for an instantly changeable truss, $m = n = 1$, $v/h = u'/a = 2u/c$.

Deflection. The coefficients in (3.11):

$C_1 = (17n^3 + 36n^2 + 25n + 6)/6$, $C_2 = ((3 + 2k)n^2 + 4n + 2)/2$,

$C_3 = ((16k^3 + 48k^2 + 50k + 3)n^2 + (36k^2 + 72k + 6)n + 30k + 3)/6$.

Forces:

$O = Pa(n + 1)/(2h)$, $D = Pc/(2h)$, $U = -Pa/h$.

Supports reactions:

$X_A = -Pan/(2h)$,

$Y_A = -P(1 + 2(k + 1)n)/2$,

$Y_B = P(1 + n(k + 1))$.

3.12.2. Load on the upper belt

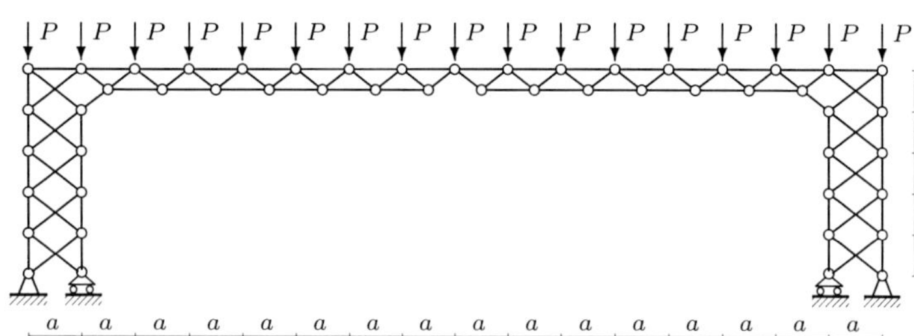

Fig. 111. Truss, $m = 4$, $n = 6$

Deflection. The coefficients in (3.11):

$C_1 = n(n+1)(7n^2+11n+3)/3,$

$C_2 = (2(3+2k)n^3 + 11n^2 + (7-8k)n + 4)/4,$

$C_3 = ((16k^3+48k^2+50k+3)n^3 + (54k^2+108k+9)n^2 -$
$-(32k^3+54k^2-44k-3)n - 36k^2 + 6k - 3)/6.$

Forces:

$O = Pa(n^2-1)/(2h)$, $D = Pc/(2h)$, $U = -Pa/h$.

Supports reactions:

$X_A = -Pa(n^2-2)/(2h),$

$Y_A = -P((2k+2)n^2 + 2n - 4k - 3)/2,$

$Y_B = P((k+1)n^2 + 2n - 2k + 1).$

Truss 3.13

A truss with a height of $(m+1)h$ consists of $8m+8n+6$ rods (Fig. 112). The total length of the rods in the truss is

$$L_{sum} = 2(m+2n+1)a + 2(m+n+2)c + 2(2m+n)h,$$

where $c = \sqrt{a^2+h^2}$.

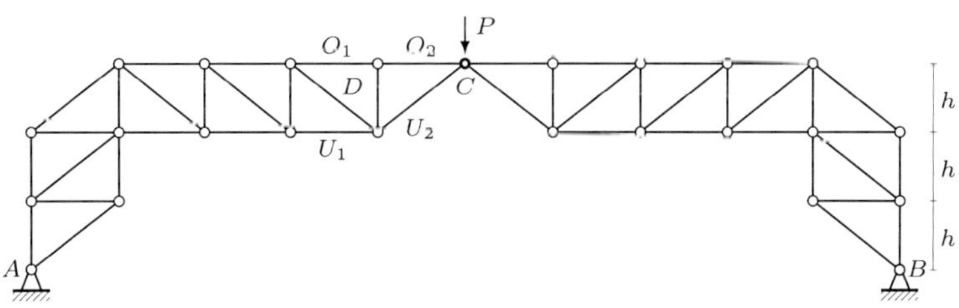

Fig. 112. Truss, $m = 2$, $n = 3$

The deflection (vertical displacement of the middle node C in the upper zone) has the form

$$\Delta = P(C_1a^3 + C_2c^3 + C_3h^3)/((m+1)^2h^2EF). \tag{3.12}$$

3.13.1. Concentrated force in the mid-span

Deflection. The coefficients in (3.12):

$C_1 = (n+2)((2n+3)(n+1)m^2 + (n+4)nm + (n+3)(2n-1))/6,$

$C_2 = (n+2)(m+1)(nm+m+1)/2,$

$C_3 = (m+1)((2n^2+5n+5)m^2 + (n^2+4n-2)m + 3n-3)/6.$

Forces:

$O_1 = O_2 = Pa(m-n-1)/(2h(m+1)),$

$D = Pc/(2h),$

$U_1 = -Pa/h,\ U_2 = -Pc/(2h).$

Supports reactions:

$X_A = Pa(n+2)/(2h(m+1)),$

$Y_A = Y_B = P/2.$

3.13.2. Load on the upper belt

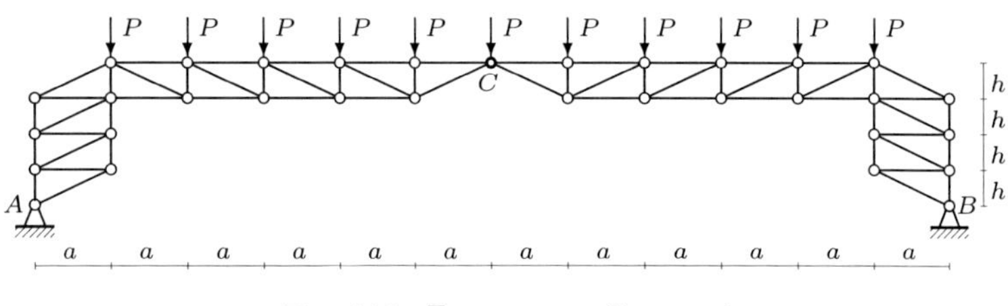

Fig. 113. Truss, $m = 3$, $n = 4$

Deflection. The coefficients in (3.12):

$C_1 = (n+2)(3(n+2)(n+1)^2m^2 + n(n^2+6n+11)m+$
$+2(2n^3+9n^2+10n-3))/12,$

$C_2 = (n+2)((n+1)^2m^2 + (n^2+3n+3)m + (n+2))/2,$

$C_3 = ((4n^3+15n^2+27n+20)m^3 + 6(n^3+4n^2+7n+1)m^2+$
$+(2n^3+15n^2+27n-32)m + 6(n+3)(n-1))/12.$

Forces:

$O_1 = O_2 = -Pa(n^2 + 4n + 3 - m)/(2h(m + 1)),$

$D = 3Pc/(2h),\ U_1 = -2Pa/h,\ U_2 = -Pc/(2h).$

Supports reactions:

$X_A = Pa(n + 2)^2/(2h(m + 1)),\ Y_A = Y_B = P(3 + 2n)/2,$

Truss 3.14

A truss with a height of $2(m + 1)h$ consists of $8m + 8n + 10$ rods (Fig. 114). The total length of the rods in the truss is

$$L_{sum} = 4(3 + 2n)a + 2(4m + 2n + 5)c + 4(2m - 1)h,$$

where $c = \sqrt{a^2 + h^2}$.

The deflection (vertical displacement of the middle node C in the upper zone) has the form

$$\Delta = P(C_1 a^3 + C_2 c^3 + C_3 h^3)/((m + 1)^2 h^2 EF). \qquad (3.13)$$

3.14.1. Concentrated force in the mid-span

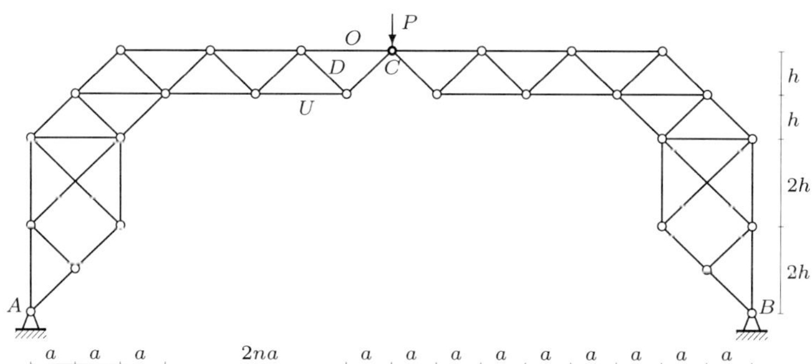

Fig. 114. Truss, $m = n = 2$

Deflection. The coefficients in (3.13):

$C_1 = (2(8n^3 + 21n^2 + 19n + 9)m^2 + 2(n + 1)(10n^2 +$

$+(3(-1)^m + 5)n - 12 + 6(-1)^m)m + 10n^3 + 3((-1)^m + 9)n^2 +$

$+2(16 + 3(-1)^m)n + 18)/6,$

$$C_2 = (m+1)((1+4m)n^2 + (9m+2)n + 3 + 5m)/2,$$

$$C_3 = (2(5+5n+2n^2)m^3 - 6nm^2 + (2n^2 + (5+3(-1)^m)n+$$
$$+6(-1)^m + 8)m + 3(n+2)((-1)^m - 1))/6.$$

Forces:

$O = Pa(m-n-1)/(2(m+1)h)$, $D = Pc/(2h)$, $U = -Pa/h$.

Supports reactions:

$X_A = Pa(n+2)/(2h(m+1))$, $Y_A = Y_B = P/2$.

3.14.2. Load on the upper belt

Deflection. The coefficients in (3.13):

$$C_1 = (6(2n^4 + 7n^3 + 10n^2 + 9n + 5)m^2 + (n+1)(14n^3+$$
$$+2(3(-1)^m + 4)n^2 + 7(3(-1)^m - 7)n - 54 + 18(-1)^m)m+$$
$$+3(n+2)(n^2+n-1)(-1)^m + 8n^4 + 31n^3 + 61n^2 + 71n + 36)/6,$$

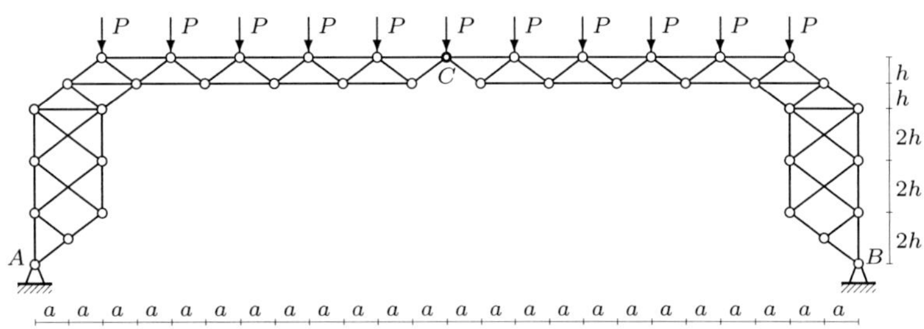

Fig. 115. Truss, $m = 3$, $n = 4$

$$C_2 = (m+1)((8m+2)n^3 + (4+25m)n^2 + (29m+10)n+$$
$$+14m + 12)/4,$$

$$C_3 = (2(4n^3 + 15n^2 + 27n + 20)m^3 - 6(3n^2 + 3n - 2)m^2+$$
$$+(4n^3 + (15 + 9(-1)^m)n^2 + 3(13 + 11(-1)^m)n+$$
$$+30(-1)^m + 38)m + 3(n+2)(3n+5)((-1)^m - 1))/12.$$

Forces:

$O = Pa(m - 3 - n^2 - 4n)/(2(m + 1)h)$,

$D = Pc/(2h)$, $U = -Pa/h$.

Supports reactions:

$X_A = Pa(n^2 + 4n + 4)/(2(m + 1)h)$, $Y_A = Y_B = P(2n + 3)/2$.

Truss 3.15

A truss with a height of $(2m+1)h$ consists of $8m+8n+10$ rods (Fig. 116). The total length of the rods in the truss is

$$L_{sum} = 8(n + 1)a + 4(m + n + 2)c + 4(2m - 1)h,$$

where $c = \sqrt{a^2 + h^2}$.

The deflection (vertical displacement of the middle node C in the upper zone) has the form

$$\Delta = P(C_1a^3 + C_2c^3 + C_3h^3)/((2m + 1)^2h^2EF). \qquad (3.14)$$

3.15.1. Concentrated force in the mid-span

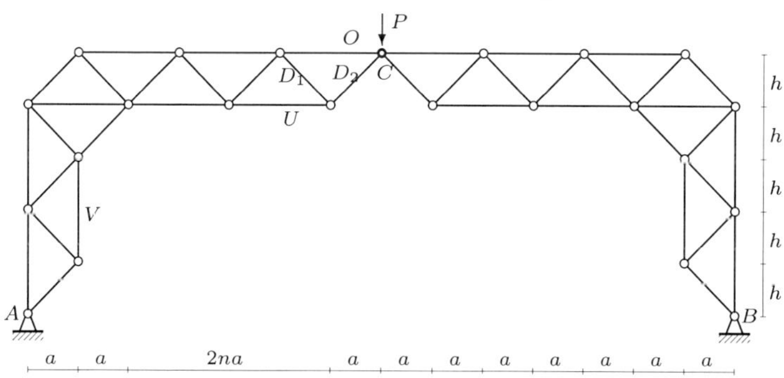

Fig. 116. Truss, $m = n = 2$

Deflection. The coefficients in (3.14):

$$C_1 = (8(4m^2 + m + 1)n^3 + 12(6m^2 - m + 3)n^2 +$$
$$+4(13m^2 - 17m + 13)n + 24(m - 1)^2)/3,$$

$$C_2 = (16m^2n^2 + 2(1 + 20m^2)n + 3 + 24m^2)/2,$$

$C_3 = 4m((4m-1)(2m-1)n^2 + 3(6m-1)(m-1)n+$
$+3(4m-1)(m-1))/3.$

Forces:

$O = Pa(m-1-n)/(h(2m+1)),$

$D_1 = Pc/(2h),\ D_2 = -Pc/(2h),\ U = -Pa/h,$

$m > 1,\ V = -P(2n+3)/(2m+1),$

$m = 1,\ V = -Pc(2n+3)/(6h).$

Supports reactions:

$X_A = Pa(2n+3)/(2h(2m+1)),\ Y_A = Y_B = P/2.$

3.15.2. Load on the upper belt

Deflection. The coefficients in (3.14):

$C_1 = (4(6m^2+m+2)n^4 + 24(3m^2-m+2)n^3+$
$+2(42m^2-65m+53)n^2 + 6(12m^2-31m+17)n+$
$+12(4m-3)(m-1))/3,$

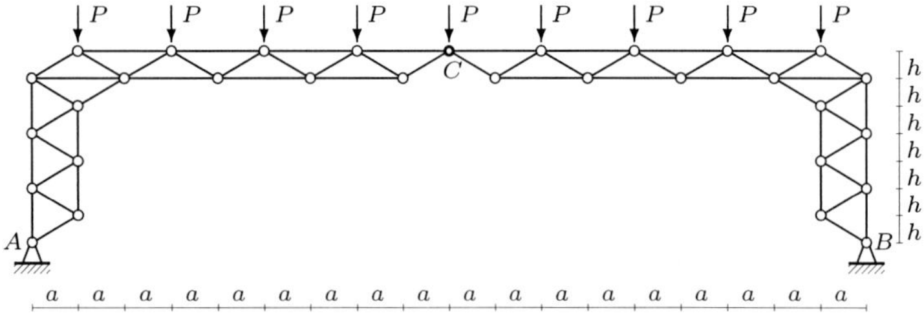

Fig. 117. Truss, $m = n = 3$

$C_2 = (16n^3m^2 + 2(28m^2-2m+1)n^2+$
$+2(36m^2-4m+3)n - 2m + 5 + 36m^2)/2,$

$C_3 = m(4(4m-1)(2m-1)n^3 + 18(6m-1)(m-1)n^2+$
$+2(5m-1)(14m-17)n + 24(3m-1)(m-1))/3.$

Forces:

$O = Pa(m - 2 - n^2 - 3n)/(h(2m + 1)),$

$D_1 = Pc/(2h),\ D_2 = -Pc/(2h),\ U = -a/h,$

$m > 1,\ V = -P(2n^2 + 6n + 5)/(2m + 1),$

$m = 1,\ V = -Pc(2n^2 + 6n + 5)/(6h).$

Supports reactions:

$X_A = Pa(2n^2 + 6n + 5)/(2h(2m + 1)),\ Y_A = Y_B = -P(2n + 3)/2.$

Truss 3.16

A truss with a height of $(m + k + 1)h$ consists of $8(m + n + k) - 2$ rods (Fig. 118). The total length of the rods in the truss is

$L_{sum} = 2(k + m + 2n - 1)a + 2(m + 2k + n + 1)c + 2(k + n + 2m - 1)h,$

where $c = \sqrt{a^2 + h^2}$.

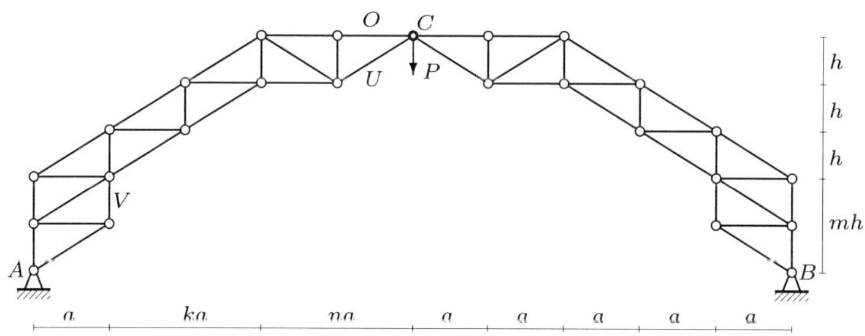

Fig. 118. Truss, $m = n = k = 2$

Case $m = n$ is considered. The deflection (vertical displacement of the middle node C in the upper zone) has the form

$$\Delta = P(C_1a^3 + C_2c^3 + C_3h^3)/(h^2EF). \qquad (3.15)$$

3.16.1. Concentrated force in the mid-span

Deflection. The coefficients in (3.15):

$C_1 = C_3 = (n - 1)(n^2 + n + 3)/3,$

$C_2 = (n^2(2k + 1) - 2nk + k + 1)/2.$

Forces:

$O = 0$, $U = -Pc/(2h)$, $V = -P/2$.

Supports reactions:

$X_A = Pa/(2h)$, $Y_A = Y_B = P/2$.

3.16.2. Load on the upper belt

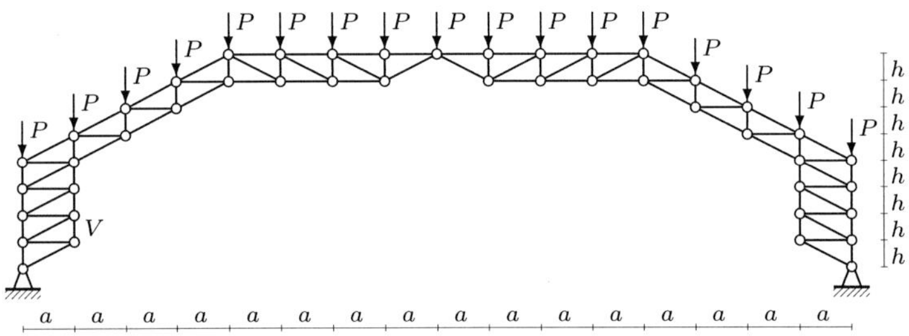

Fig. 119. Truss, $m = n = 4$, $k = 3$

Deflection. The coefficients in (3.15), $n > 1$:

$$C_1 = (n-1)(3n^3 + 4n^2 + (10+3k)n + 12)/12,$$

$$C_2 = (6(1+2k)n^3 + 6k(k-1)n^2 - (k+1)(2k^2+13k-6)n+$$
$$+(k+1)(k^2+8k+6))/12,$$

$$C_3 = (4n^4 + (4k+1)n^3 + (5-9k)n^2 + (8-k)n - 24)/12.$$

Forces:

$$O = -Pa(n+k)/(2h),$$

$$U = -Pc/(2h),$$

$$V = -P(n+k+1)/2.$$

Supports reactions:

$$X_A = Pa(n+k+1)/(2h),$$

$$Y_A = Y_B = P(2n+2k+3)/2.$$

Chapter 4

Multi-span Frames and Brackets

Truss 4.1

A truss with a height of $(2m+1)h$ consists of $16m+8n+26$ rods (Fig. 120).

The deflection (vertical displacement of the middle node C in the upper zone) has the form

$$\Delta = P(C_1a^3 + C_2c^3 + C_3h^3)/(h^2EF), \tag{4.1}$$

where $c = \sqrt{a^2+h^2}$.

The total length of the rods in the truss is

$$L_{sum} = 4(2n+5)a + 4(2m+n+5)c + 8(2m-1)h.$$

The case $m = 3$ is considered.

4.1.1. Concentrated force in the mid-span

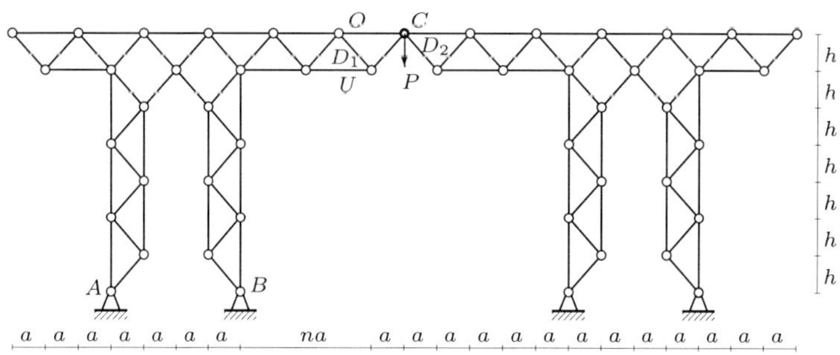

Fig. 120. Truss, $m = 3$, $n = 2$

Deflection. The coefficients in (4.1):

$C_1 = 4(68n^3 + 588n^2 + 547n + 135)/75,$

$C_2 = (1872n^2 + 1922n + 493)/50,$

$C_3 = 4(143n^2 + 170n + 53)/5.$

Forces:

$O = Pa(3 + n)/(5h),$

$D_1 = -D_2 = Pc/(2h),\ U = -Pa/h.$

Supports reactions:

$X_A = -3Pa(2n + 1)/(10h),\ Y_A = -3P(2n + 1)/10,$

$X_B = Pa(2n + 1)/(5h),\ Y_B = P(3n + 4)/5.$

4.1.2. Load on the upper belt

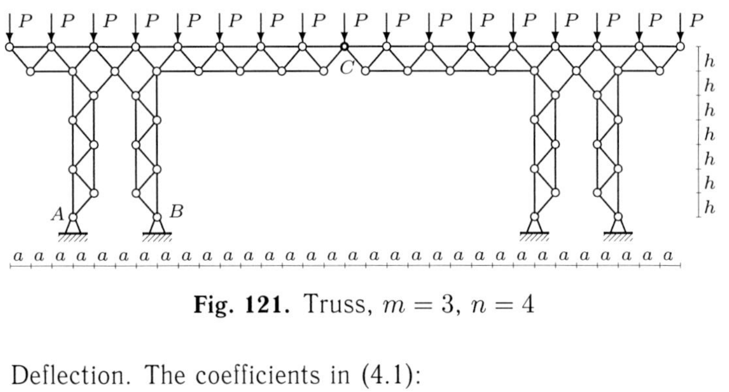

Fig. 121. Truss, $m = 3$, $n = 4$

Deflection. The coefficients in (4.1):

$C_1 = 4(53n^4 + 592n^3 + 796n^2 + 1436n + 585)/75,$

$C_2 = (1872n^3 + 2858n^2 + 5534n + 2299)/50,$

$C_3 = 4(143n^3 + 255n^2 + 525n + 241)/5.$

Forces:

$O = Pa(n^2 + n - 1)/(5h),\ D_1 = -D_2 = Pc/(2h),\ U = -Pa/h.$

Supports reactions:

$X_A = -3Pa(2n^2 + 2n + 3)/(10h),\ Y_A = -P(6n^2 + 6n - 51)/10,$

$X_B = 2Pa(n^2+n+4)/(5h)$, $Y_B = P(3n^2+8n-3)/5$.

Truss 4.2

A truss with a height of $2(m+1)h$ consists of $12m+16n+12$ rods (Fig. 122). The total length of the rods in the truss is

$$L_{sum} = 4(1+4n)a + 8(m+n+1)c + 4(3m+1)h,$$

where $c = \sqrt{a^2+h^2}$.

The deflection (vertical displacement of the middle node C in the lower zone in the left span) has the form

$$\Delta = P(C_1a^3 + C_2c^3 + C_3h^3)/(h^2\mu EF), \tag{4.2}$$

where $\mu = (16((-1)^m+1)n^2 - (-1)^m + 1)/2$.

Support A offset:

$$\delta_A = P(A_1a^3 + A_2c^3 + A_3h^3)/((2n+1)ah\mu EF). \tag{4.3}$$

4.2.1. Concentrated force in the mid-span

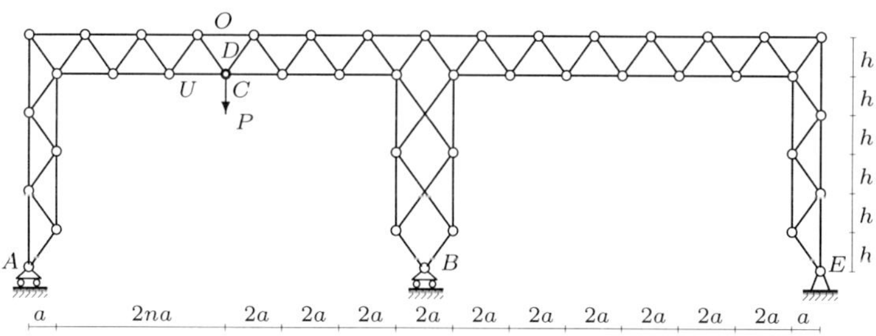

Fig. 122. Truss, $m=2$, $n=3$

Deflection. The coefficients in (4.2):

$$C_1 = (1+2n)(128(1+(-1)^m)n^4 + 32((-1)^m+1)n^3+$$
$$+8n(n+1)((-1)^m+3) + 9 + 3(-1)^m)/12,$$

$$C_2 = (64((-1)^m+1)n^3 + 16(m+2)((-1)^m+1)n^2 + 4(4(-1)^m m+$$
$$+3 + (-1)^m + 4m)n + 6 + 2(-1)^m + 5m + 3(-1)^m m)/8,$$

$$C_3 = 4(2m+1)((-1)^m+1)n^2 + 2(m-1)((-1)^m+1)n+$$

$+2m + (-1)^m m + 1.$

Support offset. The coefficients in (4.3):

$$A_1 = (96(1+2m)((-1)^m+1)n^5 + 16(1+4m)((-1)^m+1)n^4 - -(52m+22(-1)^m+76(-1)^m m+10)n^3 + (65-10(-1)^m m+ +26m+47(-1)^m)n^2 + (37m+19(-1)^m m+37(-1)^m+43)n+ +6+15m/2+6(-1)^m+9(-1)^m m/2)/3,$$

$$A_2 = (32(1+2m)((-1)^m+1)n^3 + 32(m+1)^2((-1)^m+1)n^2+ +(44(-1)^m m+32((-1)^m+1)m^2+52m+22(-1)^m+26)n+ +2(3(-1)^m+5)m^2+4(5+3(-1)^m)m+9+7(-1)^m)/4$$

$$A_3 = 16(m+1)^2((-1)^m+1)n^3 + 8(m+1)((-1)^m m - 1 + m- -(-1)^m)n^2 + (m+1)(11(-1)^m m+13m-5(-1)^m-3)n+ +(m+1)((7+5(-1)^m)m+3+(-1)^m).$$

Forces:

$O = -Pa(2n+1)(8n-(-1)^m-1)/(16hn),$

$D = Pc(8n-(-1)^m-1)/(16hn),$

$U = Pa(8n-(-1)^m-1)/(8h).$

Supports reactions:

$Y_A = P(8n-(-1)^m-1)/(16n)$, $Y_B = P(4n+(-1)^m+1)/(8n)$,

$Y_E = -P((-1)^m+1)/(16n)$, $X_E = 0$.

4.2.2. Load on the upper belt

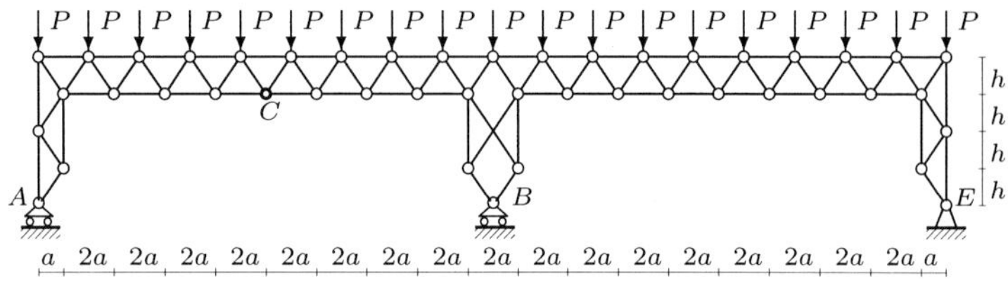

Fig. 123. Truss, $m = 1$, $n = 4$

Deflection. The coefficients in (4.2):

$$C_1 = (80((-1)^m+1)n^6 - (75+85(-1)^m)n^4 + (6(-1)^m+26)n^3 + (67+53(-1)^m)n^2 + (24(-1)^m+28)n + 3(-1)^m+3)/3,$$

$$C_2 = (64(1+(-1)^m)n^4 + 64(m+1)(1+(-1)^m)n^3 + (96(-1)^m m + 44(-1)^m + 96m + 52)n^2 + (44(-1)^m m + 36(-1)^m + 52m + 44)n + 6(-1)^m m + 7(-1)^m + 10m + 9)/8,$$

$$C_3 = (16(3m+1)(1+(-1)^m)n^3 + 48m(1+(-1)^m)n^2 + ((-1)^m(13m-11) + 19m - 5)n + 8m + 4 + 4(-1)^m m)/2.$$

Support offset. The coefficients in (4.3):

$$A_1 = (256(1+2m)((-1)^m+1)n^6 + 128(1+2m)((-1)^m+1)n^5 + 16(5-(22+26(-1)^m)m + 3(-1)^m)n^4 + 32(14-(5(-1)^m+1)m + 12(-1)^m)n^3 + 4((42(-1)^m+62)m + 95(-1)^m + 105)n^2 + 4(4(7+6(-1)^m)m + 33 + 31(-1)^m)n + 12(m+1)((-1)^m+1))/3,$$

$$A_2 = 32(2m+1)((-1)^m+1)n^4 + 32(m+1)^2((-1)^m+1)n^3 + 2(24(1+(-1)^m)m^2 + (34(-1)^m+38)m + 19(-1)^m + 21)n^2 + ((22(-1)^m+26)m^2 + 8(5(-1)^m+6)m + 23(-1)^m + 25)n + (m+1)(6(-1)^m m + 10m + 9 + 7(-1)^m)/2,$$

$$A_3 = 64(m+1)^2((-1)^m+1)n^4 + 32(m+1)(3m+1)((-1)^m+1)n^3 + 4((17+15(-1)^m)m^2 + 2(3(-1)^m+5)m - 9(-1)^m - 7)n^2 + 16((3+2(-1)^m)m^2 + ((-1)^m+3)m - (-1)^m)n + 2(m+1)(5(-1)^m m + 7m + 3 + (-1)^m).$$

Forces:

$O = -Pa(8n^3 - 4((-1)^m - 1)n^2 - ((-1)^m + 1)(4n + 1))/(8hn),$

$D = -Pc(2n + 1)((-1)^m + 1)/(8hn),$

$U = Pa(4n^2 + 2(1 - (-1)^m)n - (-1)^m - 1)/(4h).$

Supports reactions:

$Y_A = Y_E = P(8n^2 + 2(3 - (-1)^m)n - (-1)^m - 1)/(8n),$

$Y_B = P(8n^2 + 2(3 + (-1)^m)n + (-1)^m + 1)/(4n),$

$X_E = 0.$

Truss 4.3

A truss with a height of $(m+1)h$ consists of $16m+24n+21$ rods (Fig. 124). The total length of the rods in the truss is

$$L_{sum} = (12n + 5)a + 2(4m + 3n + 4)c + 8hm,$$

where $c = \sqrt{a^2 + h^2}$.

The truss is unchangeable when $m = 2, 4, 6, ...$ or when $m = 2k$, $k = 1, 2, 3, ...$. With $m = 1, 3, 5, ...$ the truss is kinematically variable (Fig. 125).

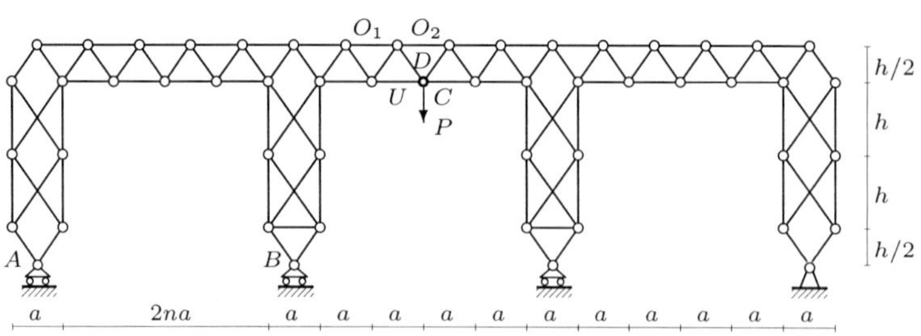

Fig. 124. Truss, $m = 2$, $n = 2$

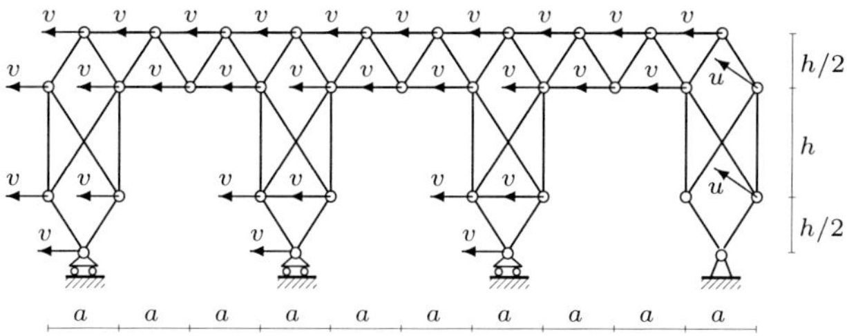

Fig. 125. Diagram of possible velocity of nodes for an instantly changeable truss, $m = 1$, $n = 1$, $u/c = v/h$.

The deflection (vertical displacement of the middle node C in the lower zone in the central span) has the form

$$\Delta = P(C_1 a^3 + C_2 c^3 + C_3 h^3)/(h^2 EF). \qquad (4.4)$$

Support A offset:

$$\delta_A = P(A_1 a^3 + A_2 c^3 + A_3 h^3)/(ahEF). \qquad (4.5)$$

4.3.1. Concentrated force in the mid-span

Deflection. The coefficients in (4.4):

$C_1 = (16n^3 + 24n^2 + 14n + 9)/12,$

$C_2 = (2k + 2n + 1)/4,\ C_3 = m/2.$

Support offset. The coefficients in (4.5):

$A_1 = -(16(2k+1)n^3 + 2(16k+9)n^2 + 2(12k+7)n - 4k - 1)/2,$

$A_2 = -k(8n-1)(2k+1),\ A_3 = -2k(4n-1)(2k+1).$

Forces:

$O_1 = -Pa(2n-1)/(2h),$

$O_2 = -Pa(2n+1)/(2h),$

$D = Pc/(2h),\ U = Pan/h.$

Supports reactions:

$Y_A = 0,\ Y_B = P/2.$

4.3.2. Load on the upper belt

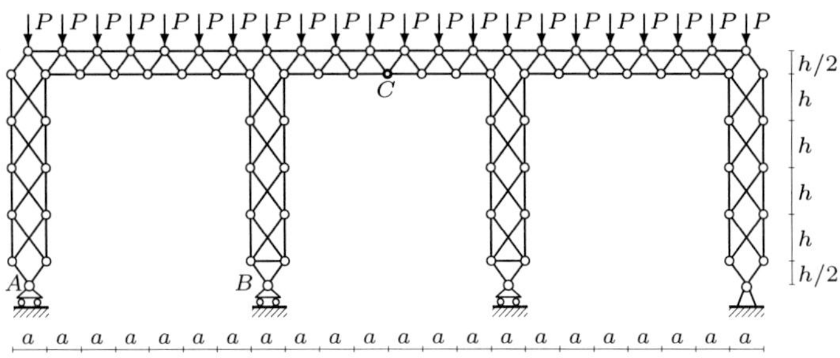

Fig. 126. Truss, $m = 4$, $n = 3$

Deflection. The coefficients in (4.4):

$C_1 = (19n^4 + 50n^3 + 59n^2 + 28n + 3)/3,$

$C_2 = (2(16k - 1)n^2 + 2(18k - 1)n + 8k - 1)/4,\ C_3 = 2nk(3 + 4n).$

Support offset. The coefficients in (4.5):

$A_1 = (256(2k + 1)n^4 + 12(88k + 47)n^3 + (472k + 314)n^2 +$

$+6(7 - 4k)n + 6 - 12k)/3,$

$A_2 = (256k(2k + 1)n^3 + 8(2k + 1)(32k + 1)n^2 +$

$+4(2k + 1)(7k + 3)n - 16k^2 - 4k + 3)/2,$

$A_3 = 4kn(32n^2 + 32n + 5)(2k + 1).$

Forces:

$O_1 = Pa(3n^2 + 5n + 3)/h,\ O_2 = Pa(n + 1)(2 + 3n)/h,$

$D = 0,\ U = -O_2.$

Supports reactions: $Y_A = 0,\ Y_B = P(2 + 3n).$

Truss 4.4

A truss with a height of $(m + 1)h$ consists of $12m + 16n + 9$ rods (Fig. 127). The total length of the rods in the truss is

$$L_{sum} = (6m + 8n + 5)c + (8n + 1)a + 6hm,$$

where $c = \sqrt{a^2 + h^2}$.

The deflection (vertical displacement of the middle node C in the lower zone in the right span) has the form

$$\Delta = P(C_1 a^3 + C_2 c^3 + C_3 h^3)/(2h^2 n^2 EF). \tag{4.6}$$

4.4.1. Concentrated force in the mid-span

Deflection. The coefficients in (4.6):

$$C_1 = n(10n^5 + 4(2(-1)^n - 3)n^4 - 16(-1)^n n^3 +$$
$$+(74 - 70(-1)^n)n^2 + (1 - (-1)^n)(23n + 4))/12,$$
$$C_2 = n(2n^2(n+2)(3 - 2(-1)^n) + (1 - (-1)^n)(n+4))/4,$$
$$C_3 = m(4n^3 + 4n^2 + (1 - (-1)^n)(n+2))/2.$$

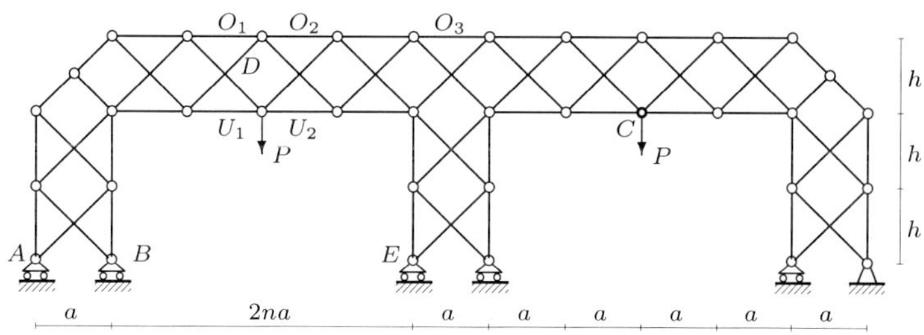

Fig. 127. Truss, $m = 2$, $n = 2$

Forces:

$$O_1 = -Pa(2n^2 + (2n-1)((-1)^n - 1))/(4hn),$$
$$O_2 = -Pa((-1)^n + 2n^2 - 1)/(4hn),$$
$$O_3 = -Pa(-1 + (-1)^n)/h,\ D = Pc(n-1)((-1)^n - 1)/(4hn),$$
$$U_1 = Pa(n-1)/(2h),\ U_2 = Pa((-1)^n - 2 + n)/(2h).$$

Supports reactions:

$$Y_A = 0,\ Y_B = P(2n - 1 + (-1)^n)/(4n),$$
$$Y_E = P(2n + 1 - (-1)^n)/(4n).$$

4.4.2. Load on the upper belt

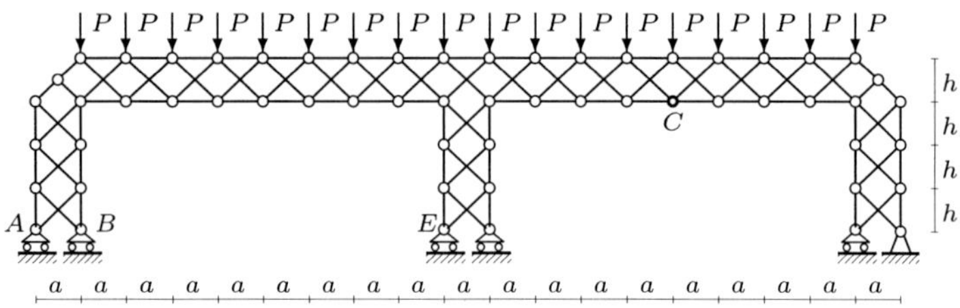

Fig. 128. Truss, $m = 3$, $n = 4$

Deflection. The coefficients in (4.6):

$$C_1 = n(10n^5 + 4(2(-1)^n - 5)n^4 + 4(3 - 7(-1)^n)n^3 + \\ +(42n^2 + 11n + 2)(1 - (-1)^n))/12,$$

$$C_2 = n(2(3 - 2(-1)^n)n^3 + (4n^2 + n + 2)(1 - (-1)^n))/4,$$

$$C_3 = m(4n^3 + 2n^2 + (1 - (-1)^n)(n + 1))/2.$$

Forces:

$$O_1 = -Pa(n + 1)(2n^2 + 2((-1)^n - 2)n + 1 - (-1)^n)/(4hn),$$

$$O_2 = -Pa(n + 1)(2n^2 - 2((-1)^n + 2)n - 1 + (-1)^n)/(4hn),$$

$$O_3 = 2Pa(n + 1)/h,$$

$$D = Pc(2(-1)^n n^2 + ((-1)^n - 1)(n - 1))/(4hn),$$

$$U_1 = Pa(2n^3 - (n + 1)(2n - 1)(1 + (-1)^n))/(4nh),$$

$$U_2 = Pa(2n^3 - (n + 1)(2n(1 - (-1)^n) + (-1)^n + 1))/(4hn).$$

Supports reactions:

$$Y_A = 0,\ Y_B = P(2n^2 - 1)/(2n),\ Y_E = P(2n^2 + 2n + 1)/(2n).$$

Truss 4.5

A truss with a height of $(2m + 1)h$ consists of $4m + 4n + 3$ rods (Fig. 129). The total length of the rods in the truss is

$$L_{sum} = 2(2n + 1)a + 2(2m + n + 1)c + 4hm,$$

where $c = \sqrt{a^2 + h^2}$.

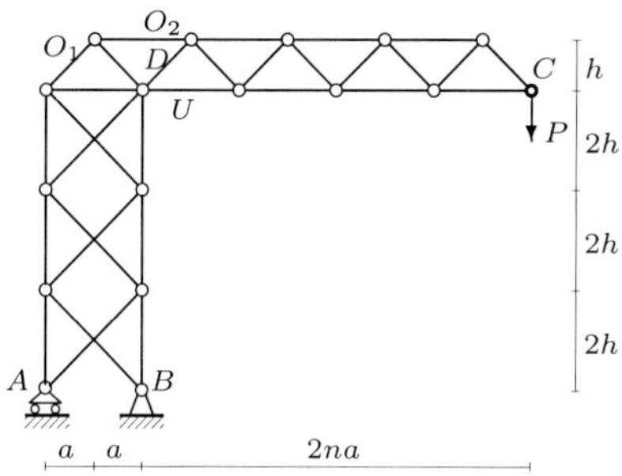

Fig. 129. Truss, $m = 3$, $n = 4$

The deflection (vertical displacement of the node C) has the form

$$\Delta = P(C_1a^3 + C_2c^3 + C_3h^2)/(h^2EF). \tag{4.7}$$

4.5.1. Concentrated force applied to the end of the console

Deflection. The coefficients in (4.7):

$C_1 = 2n(8n^2+9n+1)/3$, $C_2 = 2n(1+n)$, $C_3 = 2m(2n^2+2n+1)$.

Forces:

$O_1 = Pcn/h$, $O_2 = 2Pan/h$, $U = -Pa(2n-1)/h$, $D = -Pc/h$.

Supports reactions:

$Y_A = -Pn$, $Y_B = P(n+1)$.

4.5.2. Load on the upper belt

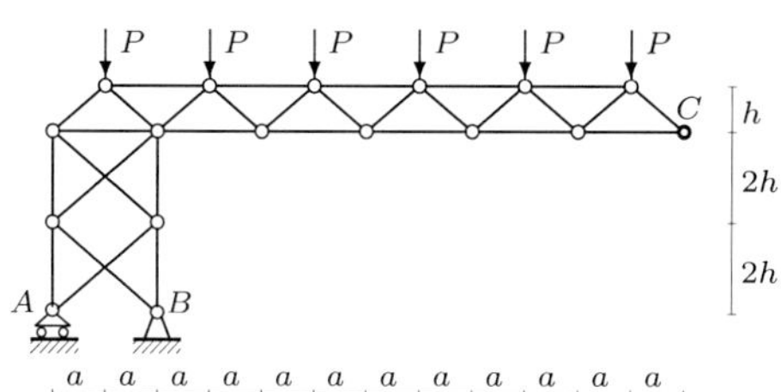

Fig. 130. Truss, $m = 2$, $n = 5$

Deflection. The coefficients in (4.7):

$C_1 = n(2n^3+3n^2-1)$, $C_2 = n^2(1+n)$, $C_3 = m(2n^3+3n^2+2n+1)$.

Forces:

$O_1 = Pc(n^2-1)/(2h)$, $O_2 = Pan^2/h$,

$U = -Pan(n-1)/h$, $D = -Pcn/h$.

Supports reactions:

$Y_A = -P(n^2-1)/2$, $Y_B = P(n+1)^2/2$.

Truss 4.6

A truss with a height of $(m-1)h+c$ consists of $4m+4n+1$ rods (Fig. 131). The deflection (vertical displacement of the node C) has the form

$$\Delta = P(C_1a^3 + C_2c^3 + C_3d^3 + C_4c^2h)/(c^2EF), \qquad (4.8)$$

where $d = \sqrt{a^2+c^2}$.

The total length of the rods in the truss is

$$L_{sum} = (m+2n+1)a + (n+2)c + (n+1)d + (m-1)(g+2h).$$

where $g = \sqrt{a^2+h^2}$.

4.6.1. Concentrated force applied to the end of the console

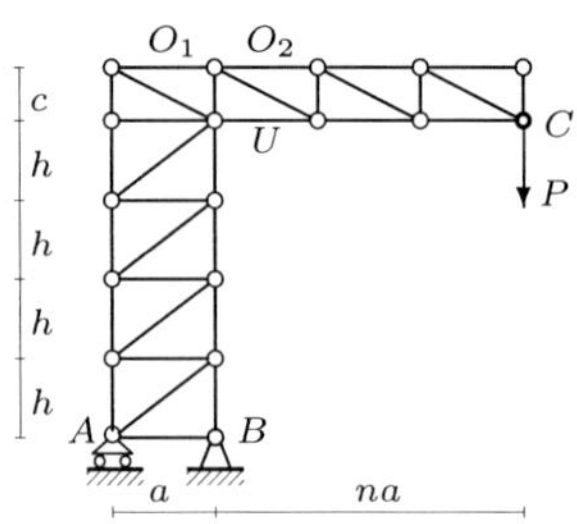

Fig. 131. Truss, $m=5$, $n=3$

Deflection. The coefficients in (4.8):

$C_1 = (2n^3 + 3n^2 + n)/3$, $C_2 = n(n+1)$, $C_3 = C_2$,

$C_4 = (1 + 2n + 2n^2)(m - 1)$.

Forces:

$O_1 = Pan/c$, $O_2 = Pa(n-1)/c$, $U = -Pan/c$.

Supports reactions: $Y_A = -nP$, $Y_B = (n+1)P$.

4.6.2. Load on the upper belt

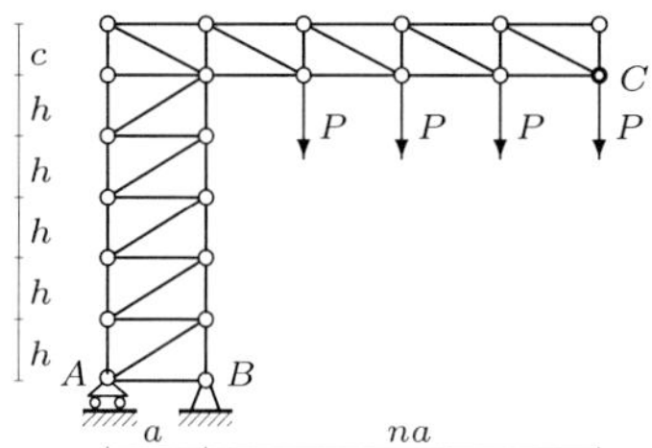

Fig. 132. Truss, $m = 6$, $n = 4$

Deflection. The coefficients in (4.8):

$C_1 = n(3n+1)(n+2)(n+1)/12$, $C_2 = n(n+1)^2/2$, $C_3 = C_2$,

$C_4 = n(2n+3)(n+1)(m-1)/2$.

Forces:

$O_1 = Pa(n+1)n/(2c)$, $O_2 = Pa(n-1)n/(2c)$,

$U = -Pa(n+1)n/(2c)$.

Supports reactions: $Y_A = -Pn(n+1)/2$, $Y_B = Pn(n+3)/2$.

Truss 4.7

A truss with a height of $mh + c$ consists of $4m + 4n + 5$ rods (Fig. 133).

The deflection (vertical displacement of the node C) has the form

$$\Delta = P(C_1a^3 + C_2c^3 + C_3d^3 + C_4c^2h)/(c^2EF), \qquad (4.9)$$

where $d = \sqrt{a^2 + c^2}$.

The total length of the rods in the truss is

$$L_{sum} = (2n+1)a + 2c + (2n+2)d + 2m(f+h),$$

where $f = \sqrt{a^2 + h^2}$.

4.7.1. Concentrated force applied to the end of the console

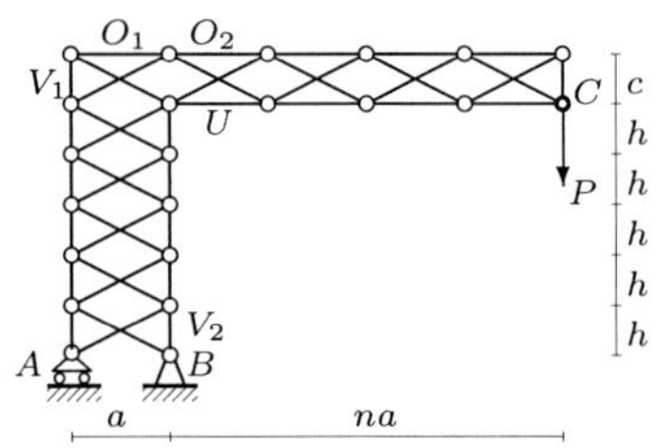

Fig. 133. Truss, $m = 5$, $n = 4$

Deflection. The coefficients in (4.9):

$C_1 = n(n+1)(2n+1)/3$, $C_2 = (1 + (-1)^n + 2n^2)/2$,

$C_3 = n(n+1)$, $C_4 = m(2n^2 + 2n + 1)$.

Forces:

$O_1 = O_2 = Pan/c$, $U_1 = -Pa(n-1)/c$,

$V_1 = Pn$, $V_2 = -P(n+1)$.

Supports reactions:

$Y_A = -nP$, $Y_B = (n+1)P$.

4.7.2. Load on the upper belt

Deflection. The coefficients in (4.9):

$C_1 = (6n^4 + 20n^3 + 24n^2 + 16n - 3(-1)^n + 3)/24$,

$C_2 = (n+2)(2n^2 - 2n + 1 + (-1)^n)/4$,

$C_3 = (4n^3 + 10n^2 + 8n - (-1)^n + 1)/8$,

$C_4 = m(n+2)(2n^2 + n + 1)/2$.

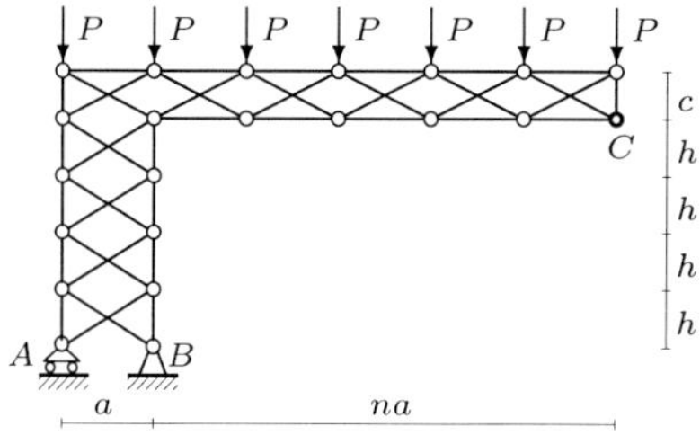

Fig. 134. Truss, $m = 4$, $n = 5$

Forces:

$O_1 = Pa(n+1)n/(2c)$, $O_2 = Pa(n^2+n+2)/(2c)$,

$U = -Pa(n+1)(n-2)/(2c)$,

$V_1 = P(n+2)(n-1)/2$, $V_2 = -P(n+1)(n+2)/2$.

Supports reactions:

$Y_A = -P(n+2)(n-1)/2$, $Y_B = P(n+1)(n+2)/2$.

Truss 4.8

A truss with a height of $2mh$ consists of $4m + 4n - 2$ rods (Fig. 135). The deflection (vertical displacement of the node C) has the form [32]

$$\Delta = P(C_1a^3 + C_2c^3 + C_3h^2)/(h^2EF), \tag{4.10}$$

where $c = \sqrt{a^2 + h^2}$.

The total length of the rods in the truss is

$$L_{sum} = 2(2n-1)a + 2(2m-1)h + 2(2m+n)c.$$

4.8.1. Concentrated force applied to the end of the console

Deflection. The coefficients in (4.10):

$C_1 = 2n(8n^2 - 18n + 13)/3$,

$C_2 = 4mn^2 - 4n(3m-1) + 9m - 2$,

$C_3 = 8(m-1)n^2 - 2n((-1)^m - 5 + 6m) - 5 + 10m + 3(-1)^m$.

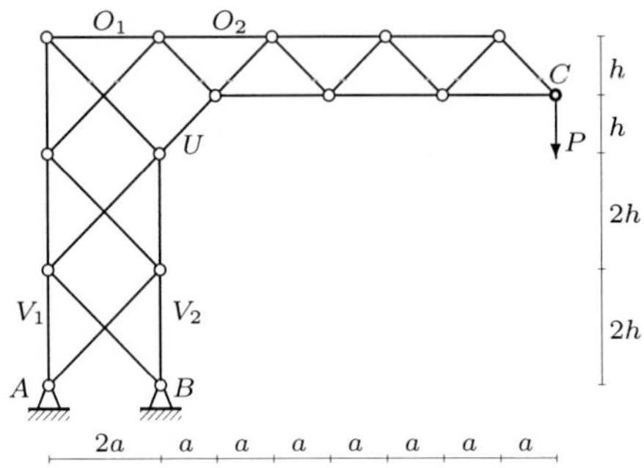

Fig. 135. Truss, $m = 3$, $n = 4$

Forces:

$O_1 = Pa/h$, $O_2 = 2a(n-1)/h$,

$U = -Pc(2n-1)/(2h)$,

$V_1 = P(2((-1)^m - 1)n - 1 - 3(-1)^m)/2$,

$V_2 = P(2((-1)^m + 1)n - 1 - 3(-1)^m)/2$.

Supports reactions:

$X_A = Pa(2n-3)(-1)^m/(2h)$, $Y_A = P(1-2n)/2$,

$X_B = -X_A$, $Y_B = P(2n+1)/2$.

4.8.2. Load on the upper belt

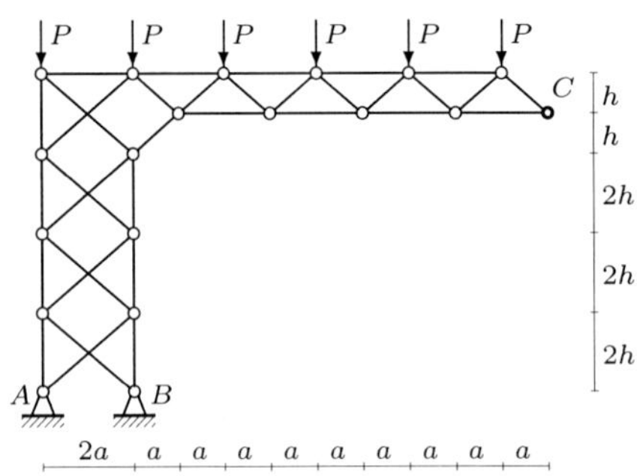

Fig. 136. Truss, $m = 4$, $n = 5$

Deflection. The coefficients in (4.10):

$$C_1 = 2n^2(n^2 - 3n + 3),$$

$$C_2 = (4mn^3 + (5 - 18m)n^2 + 2(9m - 1)n - 1)/2,$$

$$C_3 = (8(m - 1)n^3 + 3(5 - 6m - (-1)^m)n^2 + 2m+$$
$$+4(4m + (-1)^m - 2)n + 3(-1)^m - 3)/2.$$

Forces:

$$O_1 = Pan/h,\ O_2 = Pa(n - 1)^2/h,\ U = -Pcn(n - 1)/(2h),$$

$$V_1 = Pn(n(-1)^m - 3(-1)^m - 1 - n)/2,$$

$$V_2 = P(((-1)^m + 1)n^2 - (3(-1)^m + 1)n - 2)/2.$$

Supports reactions:

$$X_A = Pan(n - 3)(-1)^m/(2h),\ Y_A = P(n - n^2 + 2)/2,$$

$$X_B = -X_A,\ Y_B = Pn(n + 1)/2.$$

Truss 4.9

A truss with a height of $2(m + 1)h$ consists of $4m + 4n + 9$ rods (Fig. 137). The total length of the rods in the truss is

$$L_{sum} = (4n + 5)a + 2(3m + 2n + 3)c + (3 + 4m)h,$$

where $c = \sqrt{a^2 + h^2}$.

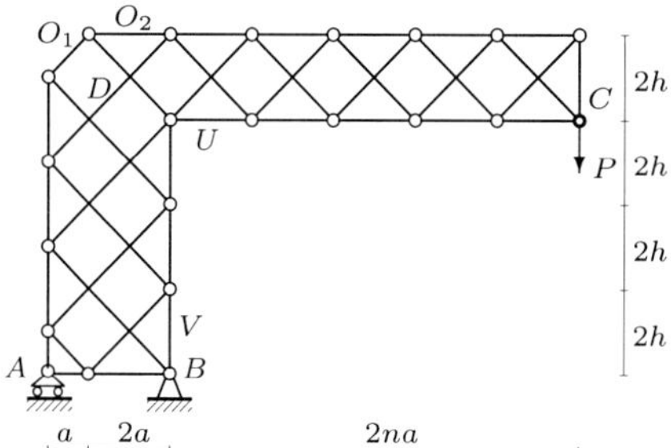

Fig. 137. Truss, $m = 3$, $n = 5$

The truss is unchangeable when $m = 1, 3, 4, 6, ...$ or when $m = (6k + (-1)^k - 1)/4$, $k = 1, 2, ...$. With $m = 2, 5, 8...$ the truss is kinematically variable (Fig. 138).

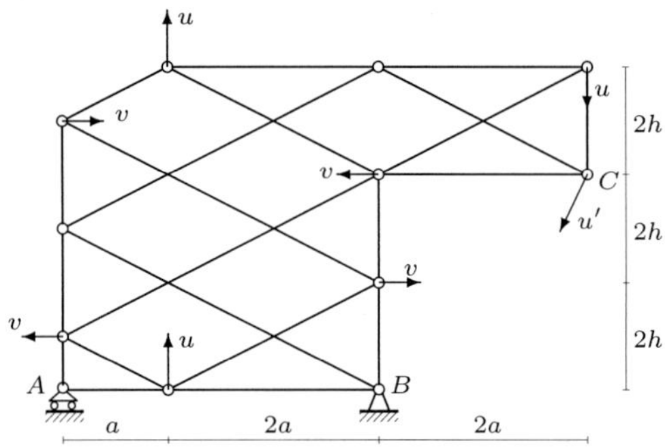

Fig. 138. Diagram of possible velocity of nodes for an instantly changeable truss, $m = 2$, $n = 1$, $v/h = u/a = u'/c$

The deflection (vertical displacement of the node C) has the form

$$\Delta = P(C_1 a^3 + C_2 c^3 + C_3 h^2)/(h^2 EF). \tag{4.11}$$

4.9.1. Concentrated force applied to the end of the console

Deflection. The coefficients in (4.11):

$C_1 = 2n(2n + 3)(4n + 1)/9$,

$C_2 = n(4n^2 + 9(2k + (-1)^k + 1)n + 18)/9$,

$C_3 = (4(20k + 2(-1)^k + 1)n^2 + 54k + 12(6k + 4(-1)^k -$
$-(-1)^{k+n} - 3)n + 9(-1)^k - 18(-1)^n + 9)/18$.

Forces:

$O_1 = Pcn/(3h)$, $O_2 = 2Pan/(3h)$,

$D = O_1$, $V = -P(4n + 3)/3$,

$U = -Pa(2n - 3)/(3h)$.

Supports reactions:

$Y_A = -2Pn/3$, $Y_B = P(2n + 3)/3$.

4.9.2. Load on the upper belt

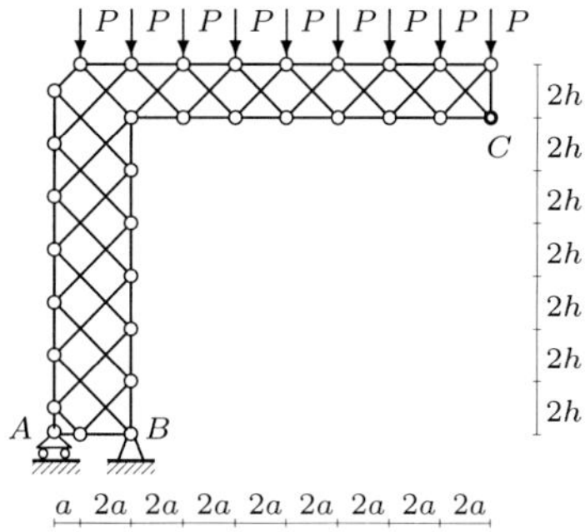

Fig. 139. Truss, $m = 6$, $n = 7$

Deflection. The coefficients in (4.11):

$$C_1 = (26n^4 + 4(3(-1)^k + 22)n^3 + 4(6(-1)^k + 19)n^2 + \\ +2(3(-1)^k - 3(-1)^{k+n} + 4)n - 9(-1)^n + 9)/36,$$

$$C_2 = (8n^4 + (36k + 30(-1)^k + 26)n^3 + (36k + 42(-1)^k + 56)n^2 + \\ +6(12(1-k) - 3(-1)^k - (-1)^{k+n})n - 9(-1)^n + 9)/36,$$

$$C_3 = ((40k + 4(-1)^k + 2)n^3 + (94k + 40(-1)^k - 3(-1)^{k+n} - 25)n^2 + \\ +72k + (64k + 58(-1)^k - 9(-1)^n - 49)n - \\ -6(-1)^k + 6(-1)^{k+n} - 9(-1)^n + 9)/18.$$

Forces:

$$O_1 = Pc(n^2 + n - 2)/(6h),\ O_2 = Pa(n^2 + n + 1)/(3h),\ D = O_1,$$

$$U = -Pa(n^2 - 2n - 2)/(3h),\ V = -P(2n^2 + 5n + 2)/3.$$

Supports reactions:

$$Y_A = -P(n^2 + n - 2)/3,\ Y_B = P(n^2 + 4n + 4)/3.$$

Truss 4.10

A truss with a height of $(m + 2)h$ consists of $4m + 4n + 13$ rods (Fig. 140). The total length of the rods in the truss is

$$L_{sum} = (3 + 2n)a + (4m + 4n + 15/2)c + (2m + 3)h,$$

where $c = \sqrt{a^2 + h^2}$

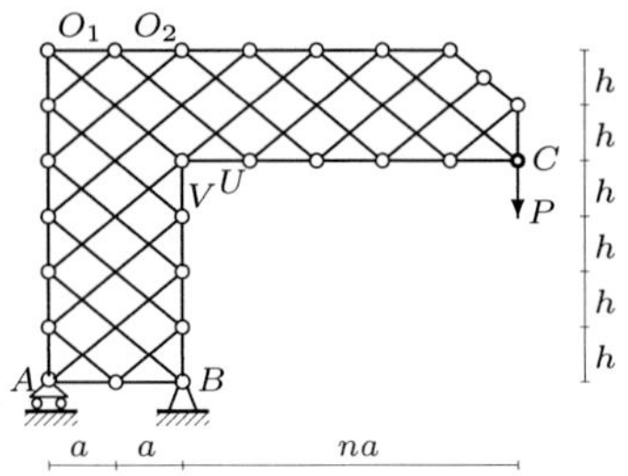

Fig. 140. Truss, $m = 4$, $n = 5$

The truss is unchangeable when $n = 2k - ((-1)^m + 1)/2$, $k = 1, 2, ...,$ $m = 1, 2, ...$, else the truss is kinematically variable (Fig. 141).

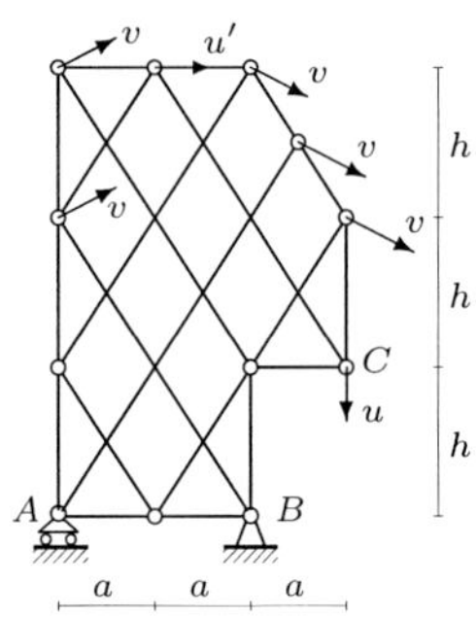

Fig. 141. Diagram of possible velocity of nodes for an instantly changeable truss, $m = n = 1$, $u/a = v/c = u'/h$

The deflection (vertical displacement of the node C) has the form

$$\Delta = P(C_1 a^3 + C_2 c^3 + C_3 h^3)/(h^2 EF). \qquad (4.12)$$

4.10.1. Concentrated force applied to the end of the console

Deflection. The coefficients in (4.12):

$C_1 = (16k^3 + 12(1 - (-1)^m)k^2 + 4(3(-1)^{m+k} - 3(-1)^m + 5-$

$$-3(-1)^k)k + 12(-1)^{m+k} - 9(-1)^m + 15 - 12(-1)^k)/12,$$

$$C_2 = (8k^2 + 2(7 - 4(-1)^k + 4(-1)^{m+k} - 5(-1)^m)k+$$
$$+(5 - 4(-1)^k - 3(-1)^m + 4(-1)^{m+k})m+$$
$$+8(1 - (-1)^k - (-1)^m + (-1)^{m+k}))/4,$$

$$C_3 = (4(m+1)k^2 + 2(1 - (-1)^m)(m + 1 - (-1)^k)k+$$
$$+(3 - (-1)^k - (-1)^m + (-1)^{m+k})m-$$
$$-4(-1)^k - 2p + 6 + 4(-1)^{m+k} - 2(-1)^m)/2,$$

where $p = \cos(m\pi/2)$.

Forces:

$$O_1 = Pa(1 - (-1)^m + (-1)^{m+k} + 4k - (-1)^k)/(4h),$$

$$O_2 = Pa(-1 + 4k + (-1)^k + (-1)^m - (-1)^{m+k})/(4h),$$

$$V = P((-1)^m - 4k - (-1)^{m+k} + (-1)^k - 5)/4,$$

$$U = -Pa(4k - (-1)^m + (-1)^{m+k} - 3 - (-1)^k)/(4h).$$

Supports reactions:

$$Y_A = -P(4k - (-1)^m - 1)/4,\ Y_B = P(4k - (-1)^m + 3)/4.$$

4.10.2. Load on the upper belt

Deflection. The coefficients in (4.12):

$$C_1 = (24k^4 + 8(1 - 3(-1)^m)k^3 + 6(10 - 4(-1)^m - 5(-1)^k+$$
$$+5(-1)^{k+m})k^2 + 2(50 - 33(-1)^k - 30(-1)^m + 27(-1)^{k+m})k+$$
$$+(1 - (-1)^k)(39 - 33(-1)^m))/24,$$

$$C_2 = (16k^3 + 4(8 - 5(-1)^k - 7(-1)^m + 5(-1)^{k+m})k^2-$$
$$-13m(-1)^k - 11m(-1)^m + 11m(-1)^{k+m} + 4(12 - 2m(-1)^m-$$
$$-3m(-1)^k + 3m(-1)^{k+m} + 4m - 10(-1)^k - 11(-1)^m+$$
$$+9(-1)^{k+m})k + 13m - 22(-1)^k - 20(-1)^m(1 - (-1)^k) + 22)/8.$$

$$C_3 = (16(m+1)k^3 + 4(3(-1)^{k+m} - 3m(-1)^m + m - 2(-1)^m-$$
$$-3(-1)^k)k^2 + 2((6 - 2(-1)^m - 3(-1)^k + 3(-1)^{k+m})m+$$

$$+12(-1)^{k+m}-14(-1)^k+7-8p-5(-1)^m)k+$$
$$+(11-7(-1)^k+5(-1)^{k+m}-5(-1)^m)m+20-10(-1)^m-$$
$$-6p+2q-18(-1)^k+12(-1)^{k+m}+2(-1)^k(p+q))/8,$$

where $q=\sin(m\pi/2)$.

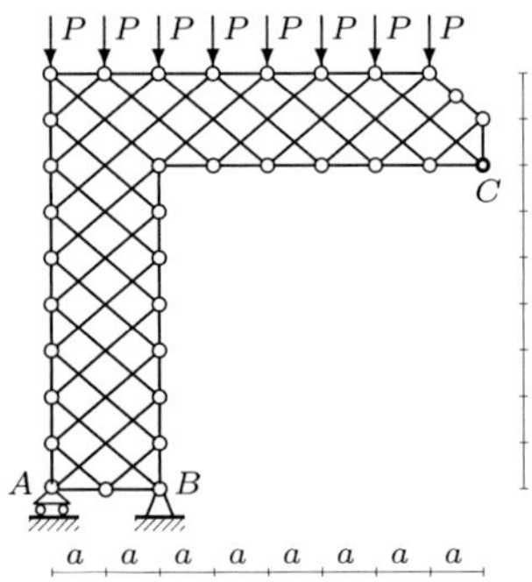

Fig. 142. Truss, $m=7$, $n=6$

Forces:

$$O_1=aP(4k^2+2((-1)^m-1)(-1)^k k-$$
$$-(-1)^m+(-1)^{m+k}-(-1)^k+1)/(4h),$$
$$O_2=aP(2k^2+((-1)^k-(-1)^{m+k}-2)k+$$
$$+2(-1)^m-(-1)^{m+k}+1)/(2h),$$
$$V=-P(4k^2+2(4-(-1)^k+(-1)^{k+m})k-3(-1)^m-$$
$$-(-1)^k+(-1)^{k+m}+3)/4,$$
$$U=-Pa(4k^2+2((-1)^{m+k}-(-1)^k-4)k+$$
$$+(-1)^m+(-1)^{m+k}-(-1)^k-1)/(4h).$$

Supports reactions:

$$Y_A=-P(4k^2-2(2+(-1)^m)k+(-1)^m-5)/4,$$
$$Y_B=P(4k^2-2(2+(-1)^m)k+4n+(-1)^m+3)/4.$$

Chapter 5

Cantilever trusses

Truss 5.1

A truss with a height of $2h$ and length na consists of $4(n+1)$ rods (Fig. 143). The total length of the rods in the truss is

$$L_{sum} = 2(an + 2cn + h),$$

where $c = \sqrt{a^2 + h^2}$.

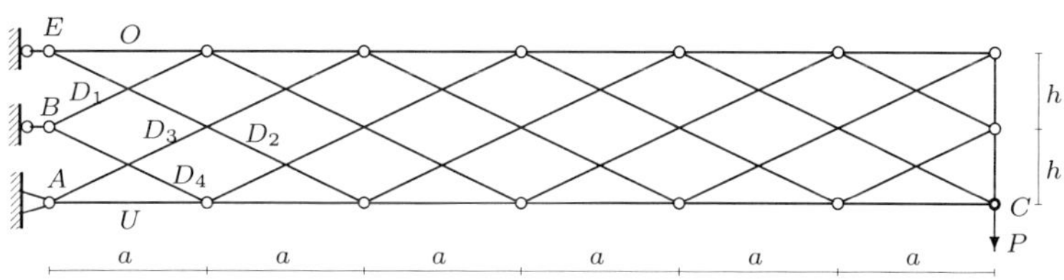

Fig. 143. Truss, $n = 6$

The deflection (vertical displacement of the node C) has the form

$$\Delta = P(C_1a^3 + C_2c^3 + C_3h^3)/(h^2EF). \tag{5.1}$$

The truss is unchangeable when $n = 2, 4, 6, ...$ or when $n = 2k$, $k = 1, 2, 3, ...$. With $n = 1, 3, 5, 7, ...$ the truss is kinematically variable (Fig. 144).

5.1.1. Concentrated force applied to the end of the console

Deflection. The coefficients in (5.1):

$C_1 = 2k(2k^2 + 1)/3$, $C_2 = 2k$, $C_3 = 1 - (-1)^k$.

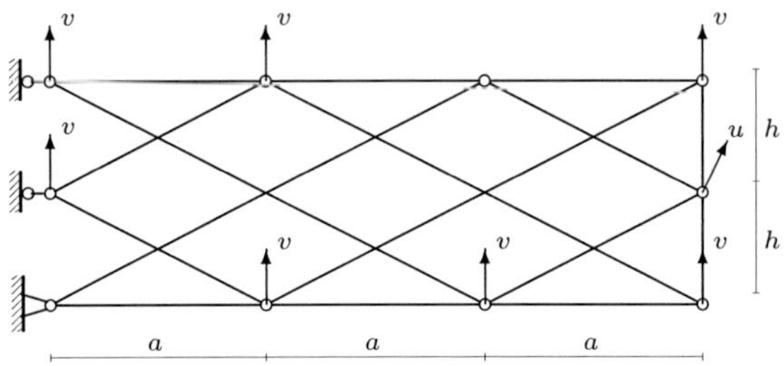

Fig. 144. Diagram of possible velocity of nodes for an instantly changeable truss, $n = 3$, $v/a = u/c$

Forces:

$O = Pak/h$, $D_1 = D_2 = D_4 = 0$, $D_3 = -Pc/h$,

$U = -P(k-1)a/h$.

Supports reactions:

$X_A = Pak/h$, $Y_A = P$, $X_B = 0$, $X_E = -Pak/h$.

5.1.2. Load on the upper belt

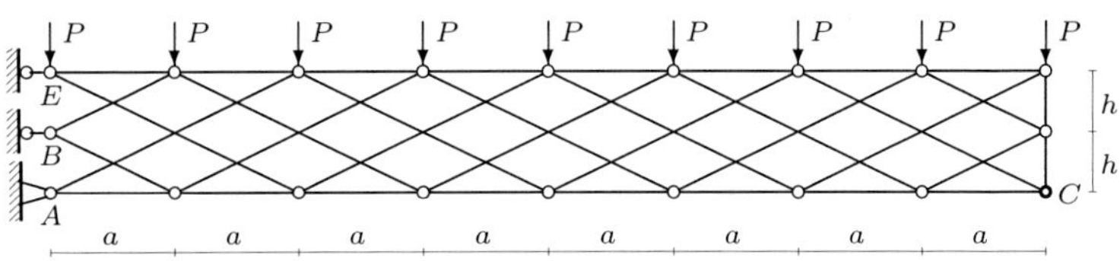

Fig. 145. Truss, $n = 2k = 8$

Deflection. The coefficients in (5.1):

$C_1 = (6k^4 + 4k^3 + 12k^2 + 8k - 3(-1)^k + 3)/6$,

$C_2 = (14k^2 + 8k - (-1)^k + 1)/4$,

$C_3 = (2k + 1)(1 - (-1)^k)/2$.

Forces:

$O = Pa(4k^2 + 2k - (-1)^k + 5)/(4h)$, $D_1 = Pc((-1)^k - 1)/(4h)$,

$D_2 = -Pc/h$, $D_3 = -Pc(1+2k)/h$,

$D_4 = D_1$, $U = -Pa(4k^2 - 6k + (-1)^k - 5)/(4h)$.

Supports reactions:

$X_A = Pa(4k^2+2k+(-1)^k-1)/(4h)$, $X_B = -Pa((-1)^k-1)/(2h)$,

$X_E = -Pa(4k^2 + 2k - (-1)^k + 1)/(4h)$, $Y_A = P(1+2k)$.

Truss 5.2

A truss with a height of $2h$ and length na consists of $4n+3$ rods (Fig. 146). The total length of the rods in the truss is

$$L_{sum} = 2an + 2cn + dn + 2h,$$

where $c = \sqrt{a^2+h^2}$, $d = \sqrt{a^2+4h^2}$.

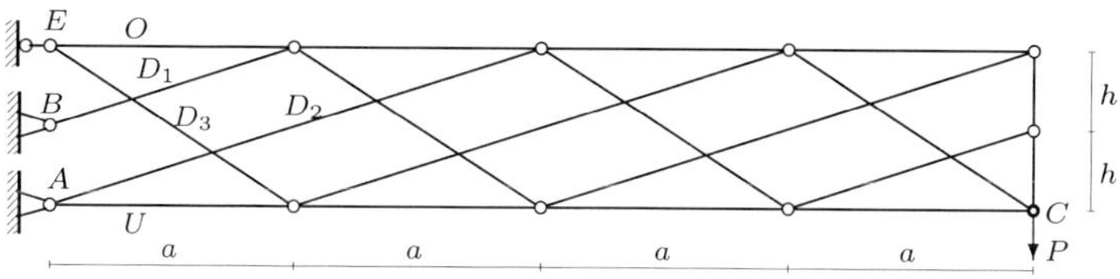

Fig. 146. Truss, $n = 4$

The deflection (vertical displacement of the node C) has the form

$$\Delta = P(C_1a^3 + C_2c^3 + C_3h^3 + C_4d^3)/(h^2EF). \qquad (5.2)$$

The truss is unchangeable when $n = 1, 3, 4, 6, 7...$ or when $n = ((-1)^k + 6k - 1)/4$, $k = 1, 2, 3, ...$. With $n = 2, 5, 8, 11, ...$ the truss is kinematically variable (Fig. 147).

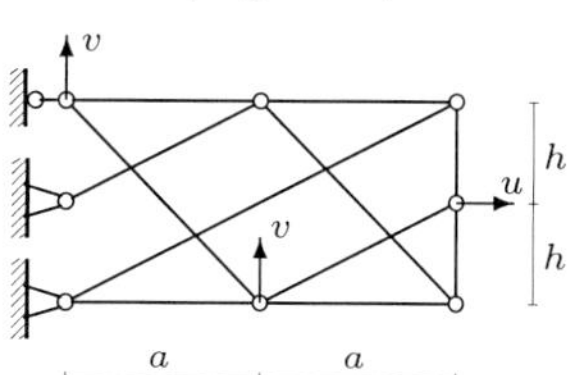

Fig. 147. Diagram of possible velocity of nodes for an instantly changeable truss, $n = 2$, $v/a = u/h$

5.2.1. Concentrated force applied to the end of the console

Deflection. The coefficients in (5.2):

$C_1 = (18k^3 + 9k^2(-1)^k - 9k^2 - 15k(-1)^k + 27k + 5(-1)^k - 5)/32,$

$C_2 = k$, $C_3 = 1 - (-1)^k$, $C_4 = (2k + (-1)^k - 1)/16.$

Forces:

$O = P(6k - (-1)^k + 1)a/(8h)$, $D_1 = P((-1)^k - 1)c/(2h)$,

$D_2 = -P((-1)^k + 1)c/(2h)$, $D_3 = 0$, $U = P(7 - 6k + (-1)^k)a/(8h)$.

Supports reactions:

$X_A = 3P(2k + (-1)^k - 1)a/(8h)$, $Y_A = P((-1)^k + 1)/2$,

$X_B = P(1 - (-1)^k)a/(2h)$, $Y_B = P(1 - (-1)^k)/2$,

$X_E = P((-1)^k - 1 - 6k)a/(8h)$.

5.2.2. Load on the upper belt

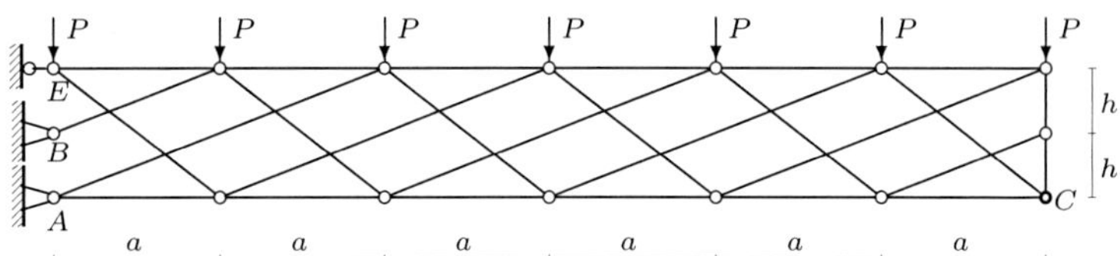

Fig. 148. Truss, $n = 6$

Deflection. The coefficients in (5.2):

$C_1 = (162k^4 + 36(3(-1)^k + 1)k^3 + 6(61 - 17(-1)^k)k^2 +$

$\quad + 20(25 - (-1)^k)k - 27 + 27(-1)^k)/512,$

$C_2 = (6k^2 + 2((-1)^k + 5)k + 1 - (-1)^k)/8,$

$C_3 = (k + 1)(1 - (-1)^k)/2,$

$C_4 = (6k^2 + 2(1 + (-1)^k)k + 3(-1)^k - 3)/64.$

Forces:

$O = Pa(18k^2 + 2k(-1)^k + 18k - 7(-1)^k + 23)/(32h),$

$D_1 = P(k(-1)^k - 3k + 2(-1)^k - 2)c/(4h),$

$D_2 = -P(k(-1)^k + 3k + 3(-1)^k + 1)c/(4h),\ D_3 = -Pd/(2h),$

$U = -Pa(18k^2 + 2k(-1)^k - 30k - 15(-1)^k - 17)/(32h),$

Supports reactions:

$X_A = Pa(18k^2 + 2(5(-1)^k - 3)k + 9(-1)^k - 9)/(32h),$

$Y_A = P((3 + (-1)^k)k + 1 + 3(-1)^k)/4,$

$X_B = Pa((3 - (-1)^k)k - 2(-1)^k + 2)/(4h),$

$Y_B = P((3 - (-1)^k)k + 2(1 - (-1)^k))/4,$

$X_E = -Pa(18k^2 + 2((-1)^k + 9)k + 7 - 7(-1)^k)/(32h).$

Truss 5.3

A truss with a height of $3h$ and length $2na$ consists of $4n+3$ rods (Fig. 149). The total length of the rods in the truss is

$$L_{sum} = (4n - 1)a + 6cn + 2h,$$

where $c = \sqrt{a^2 + h^2}$.

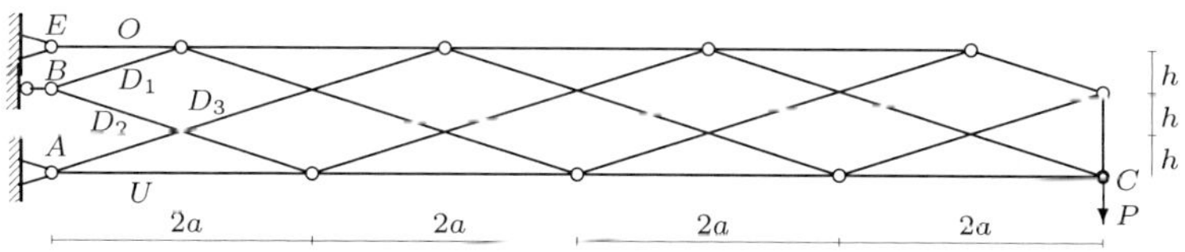

Fig. 149. Truss, $n = 4$

The deflection (vertical displacement of the node C) has the form

$$\Delta = P(C_1a^3 + C_2c^3 + C_3h^3)/(h^2EF). \qquad (5.3)$$

5.3.1. Concentrated force applied to the end of the console

Deflection. The coefficients in (5.3):

$C_1 = 2(24n^3 + 3(41 - 32\cos\varphi)n + 32\sqrt{3}\sin\varphi)/81,$

$C_2 = 2((7 - 4\cos\varphi)n)/3,\ C_3 = 16(1 - \cos\varphi)/3,$

where $\varphi = 2n\pi/3$.

Forces:

$O = 2aP(3n + 4\sqrt{3}\,\sin\varphi)/(9h)$, $D_1 = -2Pc(\sqrt{3}\,/(3h))\sin\varphi$,

$D_2 = -2cP(\sqrt{3}/(3h))\,\sin\varphi$, $D_3 = -Pc/h$,

$U = Pa(9 - 6n + 4\sqrt{3}\,\sin\varphi)/(9h)$.

Supports reactions:

$X_A = 2aP(3n - 2\sqrt{3}\,\sin\varphi)/(9h)$, $Y_A = P$,

$X_E = -2aP(3n + 4\sqrt{3}\,\sin\varphi)/(9h)$, $Y_E = 0$,

$X_B = 4aP(\sqrt{3}/(3h))\sin\varphi$,

5.3.2. Load on the upper belt

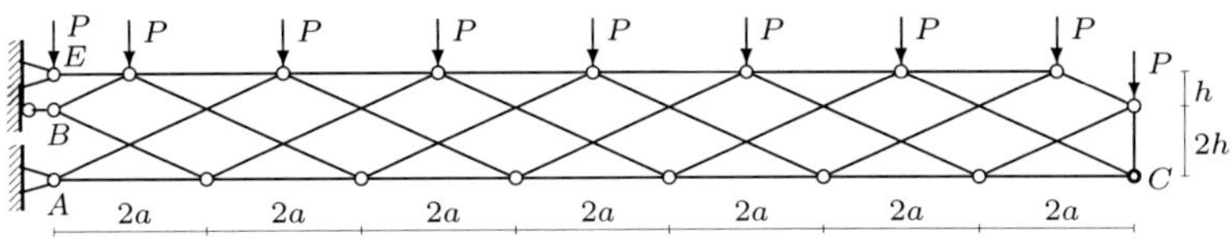

Fig. 150. Truss, $n = 7$

Deflection. The coefficients in (5.3):

$$C_1 = 2(9n^4 + 24n^3 + 12(11 - 8\cos\varphi)n^2 +$$
$$+3(41 + 32\sqrt{3}\,\sin\varphi - 32\cos\varphi)n + 32\sqrt{3}\,\sin\varphi +,$$
$$+48 - 48\cos\varphi)/81,$$

$$C_2 = (3(13 - 8\cos\varphi)n^2 + 2(21 - 12\cos\varphi + 8\sqrt{3}\,\sin\varphi)n -$$
$$-4\cos\varphi + 4)/9,$$

$$C_3 = 8(4(1 - \cos\varphi)n + 3 + 2\sqrt{3}\,\sin\varphi - 3\cos\varphi)/9.$$

Forces:

$O = Pa(3n^2 + (8\sqrt{3}\,\sin\varphi + 6)n + 8\sqrt{3}\,\sin\varphi + 8 - 8\cos\varphi)/(9h)$,

$D_1 = D_2 = -2cP(\sqrt{3}\;n\sin\varphi + \sqrt{3}\,\sin\varphi - \cos\varphi + 1)/(3h)$,

$D_3 = -P(n+1)c/h,$

$U = -Pa(3n^2 - (3+4\sqrt{3}\,\sin\varphi)n + 4\cos\varphi - 13 - 4\sqrt{3}\,\sin\varphi)/(9h).$

Supports reactions:

$X_A = Pa(3n^2 + (6 - 4\sqrt{3}\,\sin\varphi)n + 4\cos\varphi - 4 - 4\sqrt{3}\,\sin\varphi)/(9h),$

$Y_A = P(n+1),$

$X_E = -Pa(3n^2 + (6+8\sqrt{3}\,\sin\varphi)n - 8\cos\varphi + 8 + 8\sqrt{3}\,\sin\varphi)/(9h),$

$Y_E = P,$

$X_B = 4aP(\sqrt{3}\,(n+1)\sin\varphi - \cos\varphi + 1)/(3h).$

Truss 5.4

A truss with a height of $2h$ and length na consists of $10n$ rods (Fig. 151). The total length of the rods in the truss is

$$L_{sum} = n(4a + 2c + 2d + 3h),$$

where $c = \sqrt{a^2 + h^2}$, $d = \sqrt{a^2 + 4h^2}$.

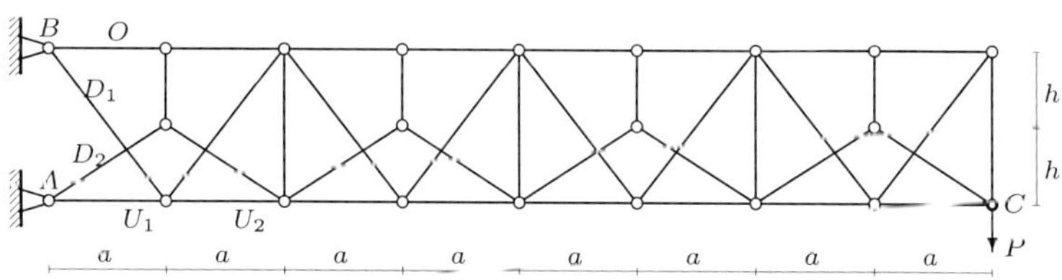

Fig. 151. Truss, $n = 4$

The deflection (vertical displacement of the node C) has the form

$$\Delta = P(C_1 a^3 + C_2 d^3 + C_3 h^3)/(h^2 EF). \quad (5.4)$$

5.4.1. Concentrated force applied to the end of the console

Deflection. The coefficients in (5.4):

$C_1 = n(8n^2 + 1)/6$, $C_2 = n/2$, $C_3 = 2$.

Forces:

$O = P(2n-1)a/(2h)$, $D_1 = Pd/(2h)$, $D_2 = 0$,

$U_1 = -Pna/h$, $U_2 = -P(n-1)a/h$.

Supports reactions:

$X_B = -Pna/h$, $Y_B = P$, $X_A = Pna/h$, $Y_A = 0$.

5.4.2. Load on the upper belt

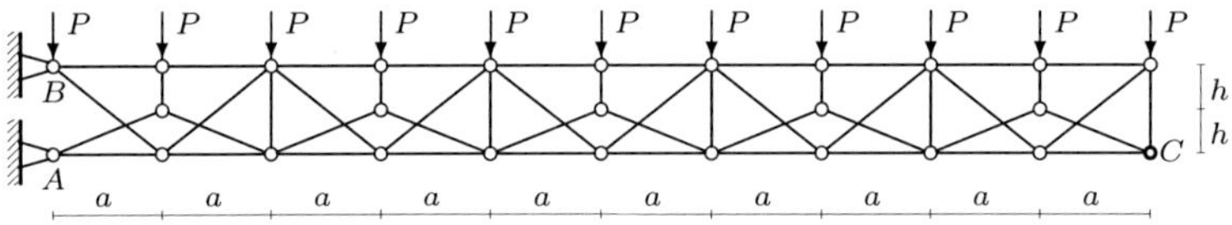

Fig. 152. Truss, $n = 5$

Deflection. The coefficients in (5.4):

$C_1 = n(12n^3 + 8n^2 + 1)/12$, $C_2 = n(1 + 2n)/4$, $C_3 = 1$.

Forces:

$O = P(4n^2 - 2n + 1)a/(4h)$,

$D_1 = P(4n-1)d/(4h)$, $D_2 = -Pc/(2h)$,

$U_1 = -P(n+1)(2n-1)a/(2h)$, $U_2 = -Pn(2n-3)a/(2h)$.

Supports reactions:

$X_A = Pn(2n+1)a/(2h)$, $Y_A = P/2$,

$X_B = -Pn(2n+1)a/(2h)$, $Y_B = P(4n+1)/2$.

Truss 5.5

A truss with a height of h and length $2na$ consists of $4n$ rods (Fig. 153). The total length of the rods in the truss is

$$L_{sum} = (4n-1)a + 2nc,$$

where $c = \sqrt{a^2 + h^2}$.

The deflection (vertical displacement of the node C) has the form

$$\Delta = P(C_1 a^3 + C_2 c^3)/(h^2 EF). \tag{5.5}$$

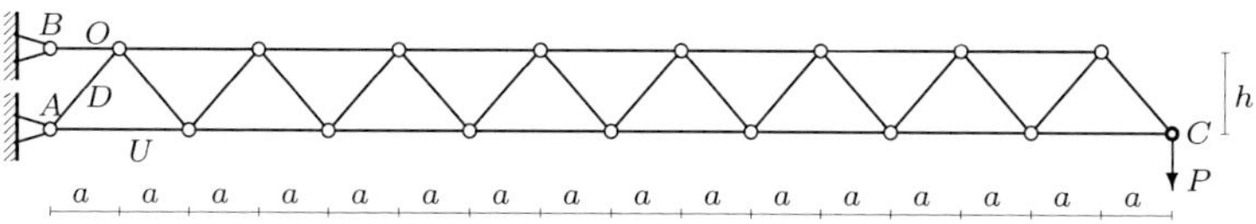

Fig. 153. Truss, $n = 8$

5.5.1. Concentrated force applied to the end of the console

Deflection. The coefficients in (5.5):

$C_1 = 2n(8n^2 + 1)/3$, $C_2 = 2n$.

Forces:

$O = 2naP/h$, $D = -Pc/h$,

$U = -P(2n - 1)a/h$.

Supports reactions:

$X_A = 2naP/h$, $Y_A = P$,

$X_B = -2naP/h$, $Y_B = 0$.

5.5.2. Load on the upper belt

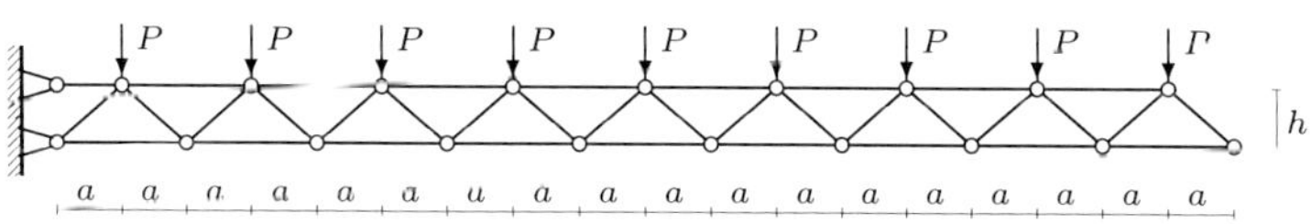

Fig. 154. Truss, $n = 9$

Deflection. The coefficients in (5.5):

$C_1 = 2n^4$, $C_2 = n^2$.

Forces:

$O = Pn^2 a/h$, $D = -Pcn/h$, $U = -P(n - 1)na/h$.

Supports reactions: $X_A = Pan^2/h$, $Y_A = Pn$, $X_B = -X_A$, $Y_B = 0$.

Truss 5.6

A truss with a height of $2nh$ and length $2na$ consists of $10n - 3$ rods (Fig. 155). The total length of the rods in the truss is

$$L_{sum} = 2an + n(1 + 2n)(h + c),$$

where $c = \sqrt{a^2 + h^2}$.

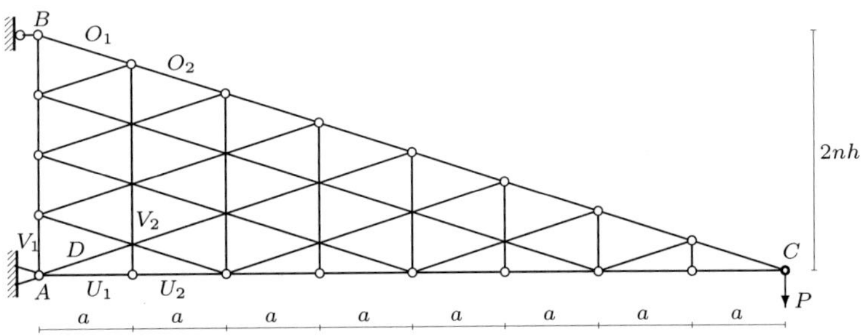

Fig. 155. Truss, $n = 4$

The deflection (vertical displacement of the node C) has the form

$$\Delta = P(C_1a^3 + C_2c^3 + C_3h^3)/(h^2EF). \quad (5.6)$$

5.6.1. Concentrated force applied to the end of the console

Deflection. The coefficients in (5.6):

$C_1 = C_2 = C_3 = 2n$, $n > 1$.

Forces:

$O_1 = Pc/h$, $O_2 = Pc/h$, $D = 0$,

$U_1 = U_2 = -Pa/h$, $V_1 = -P$, $V_2 = 0$.

Supports reactions:

$X_A = Pa/h$, $Y_A = P$, $X_B = -Pa/h$.

5.6.2. Load on the upper belt

Deflection. The coefficients in (5.6):

$C_1 = n(n + 1)$, $C_2 = n(2n + 3)/2$, $C_3 = n(3n + 2)$, $n > 1$.

Forces:

$O_1 = Pc(2n+1)/(2h)$, $O_2 = Pcn/h$, $D = -Pc/(2h)$,

$V_1 = -P(4n+1)/2$, $V_2 = 0$,

$U_1 = U_2 = -2Pa/h$.

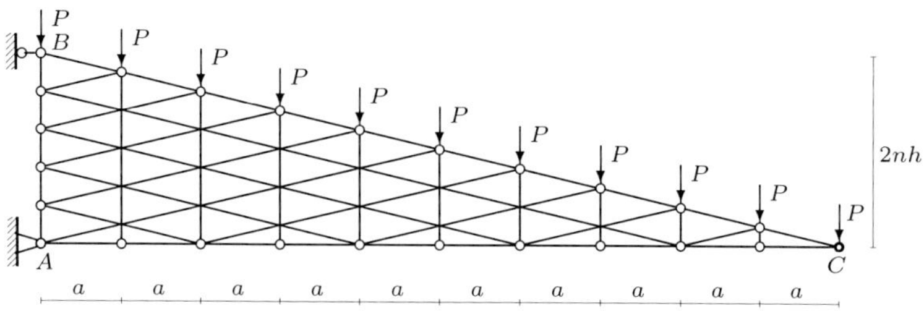

Fig. 156. Truss, $n = 5$

Supports reactions:

$X_A = Pa(2n+1)/(2h)$, $Y_A = P(2n+1)$, $X_B = -X_A$.

Truss 5.7

A truss with a height of h and length na consists of $4n+1$ rods (Fig. 157). The total length of the rods in the truss is

$$L_{sum} = (n+1)h + n(2a+c),$$

where $c = \sqrt{a^2+h^2}$.

The deflection (vertical displacement of the node C) has the form

$$\Delta = P(C_1a^3 + C_2c^3 + C_3h^3)/(h^2EF). \qquad (5.7)$$

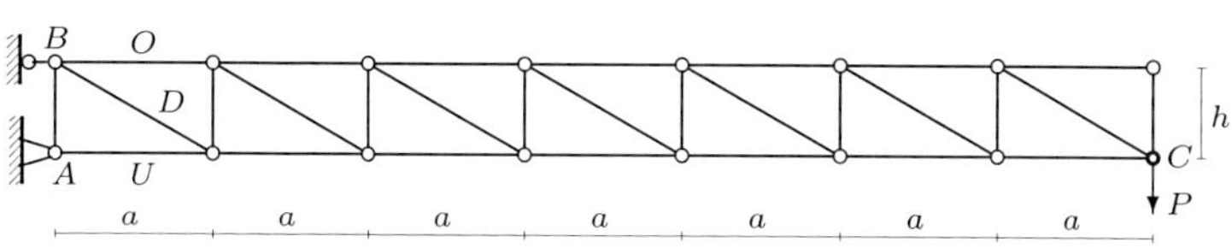

Fig. 157. Truss, $n = 7$

5.7.1. Concentrated force applied to the end of the console

Deflection. The coefficients in (5.7):

$C_1 = n(1 + 2n^2)/3$, $C_2 = C_3 = n$.

Forces:

$O = Pa(n - 1)/h$, $D = Pc/h$, $U = -Pna/h$.

Supports reactions:

$X_A = Pna/h$, $X_B = -Pna/h$, $Y_A = P$.

5.7.2. Load on the upper belt

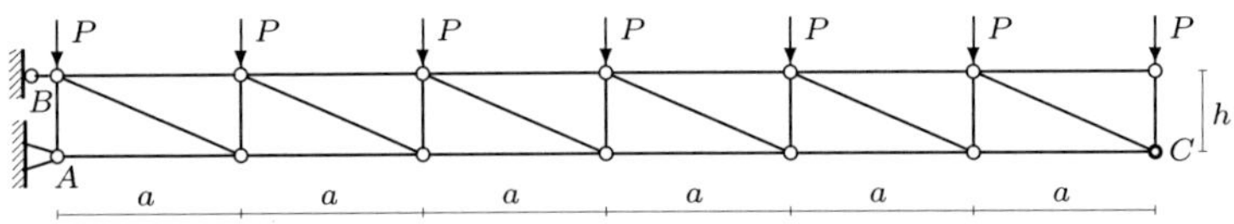

Fig. 158. Truss, $n = 6$

Deflection. The coefficients in (5.7):

$C_1 = n(n+1)(3n^2+n+2)/12$, $C_2 = n(n+1)/2$, $C_3 = n(n+3)/2$.

Forces:

$O = Pan(n - 1)/(2h)$, $D = Pcn/h$,

$U = -Pan(n + 1)/(2h)$.

Supports reactions:

$X_A = -X_B = Pa(n + 1)n/(2h)$, $Y_A = P(n + 1)$.

Truss 5.8

A truss with a height of h and length na consists of $4n + 1$ rods (Fig. 159). The total length of the rods in the truss is

$$L_{sum} = 2an + 2cn + h,$$

where $c = \sqrt{a^2 + h^2}$.

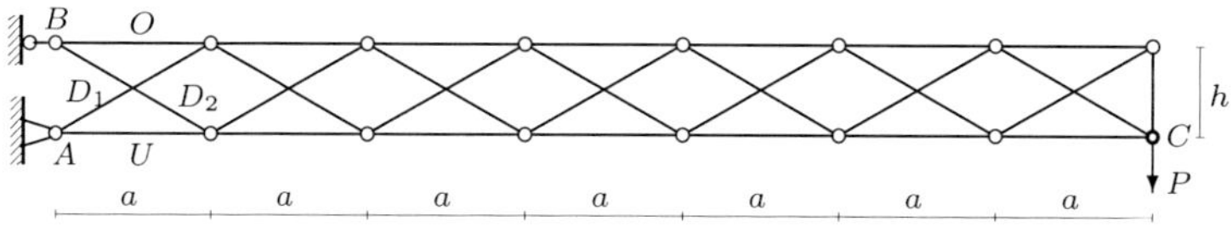

Fig. 159. Truss, $n = 7$

The deflection (vertical displacement of the node C) has the form

$$\Delta = P(C_1a^3 + C_2c^3 + C_3h^3)/(h^2EF). \tag{5.8}$$

5.8.1. Concentrated force applied to the end of the console

Deflection. The coefficients in (5.8):

$C_1 = n(2n^2 + 1)/3$, $C_2 = n$, $C_3 = (1 - (-1)^n)/2$.

Forces:

$O = Pan/h$, $D_1 = -Pc/h$, $D_2 = 0$, $U = -P(n - 1)a/h$.

Supports reactions:

$X_A = Pna/h$, $X_B = -Pna/h$, $Y_A = P$.

5.8.2. Load on the upper belt

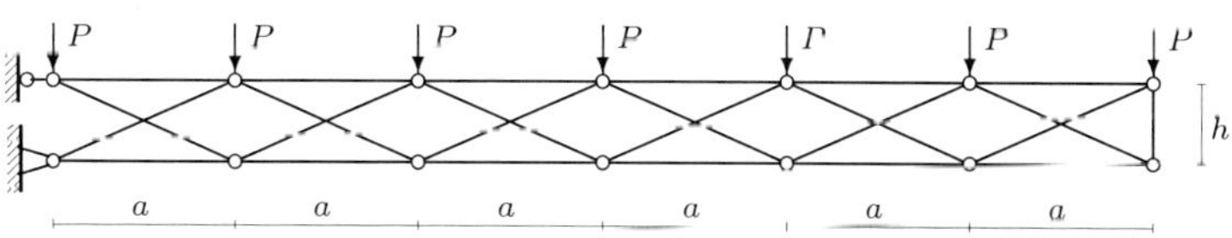

Fig. 160. Truss, $n = 6$

Deflection. The coefficients in (5.8):

$$C_1 = (6n^4 + 8n^3 + 12n^2 + 16n - 3(-1)^n + 3)/24, \tag{5.9}$$

$C_2 = (6n^2 + 8n - (-1)^n + 1)/8$, $C_3 = (n + 1)(1 - (-1)^n)/4$.

Forces:

$O = P(n^2 + n + 2)a/(2h)$,

$D_1 = -Pc(n+1)/h$, $D_2 = -Pc/h$,

$U = -Pa(n+1)(n-2)/(2h)$.

Supports reactions:

$X_A = Pa(n+1)n/(2h)$, $X_B = -X_A$,

$Y_A = P(n+1)$.

Truss 5.9

A truss with a height of $2h$ and length na consists of $6n+1$ rods (Fig. 161). The total length of the rods in the truss is

$$L_{sum} = 2an + 2cn + 2(n+1)h,$$

where $c = \sqrt{a^2 + h^2}$.

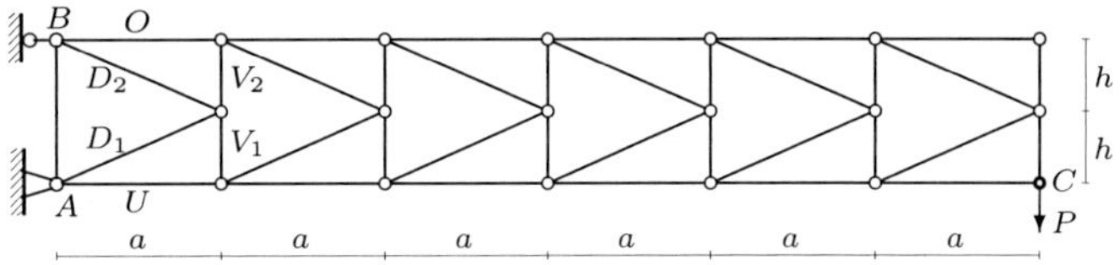

Fig. 161. Truss, $n = 6$

The deflection (vertical displacement of the node C) has the form

$$\Delta = P(C_1a^3 + C_2c^3 + C_3h^3)/(h^2EF). \tag{5.10}$$

5.9.1. Concentrated force applied to the end of the console

Deflection. The coefficients in (5.10):

$C_1 = n(2n-1)(n-1)/12$, $C_2 = n/2$, $C_3 = 1 + n/2$.

Forces ($n > 1$):

$O = P(n-1)a/h$, $D_1 = -Pc/(2h)$,

$D_2 = -D_1$, $U = -P(n-1)a/h$,

$V_1 = P/2$, $V_2 = -P/2$.

Supports reactions: $X_A = Pan/(2h)$, $X_B = -X_A$, $Y_A = P$.

5.9.2. Load on the upper belt

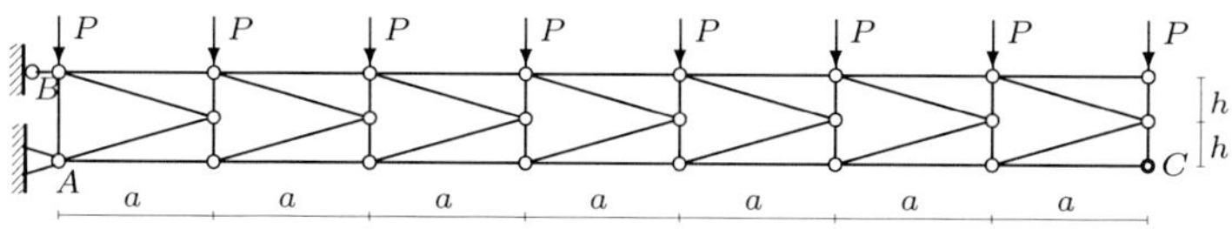

Fig. 162. Truss, $n = 7$

Deflection. The coefficients in (5.10):

$C_1 = n(n^2 - 1)(3n - 2)/48$, $C_2 = n(n + 1)/4$,

$C_3 = (n + 1)(n + 2)/4$.

Forces ($n > 1$):

$O = Pa(n - 1)n/(4h)$, $D_1 = -Pcn/(2h)$, $D_2 = -D_1$,

$V_1 = P(n - 1)/2$, $V_2 = -P(n + 1)/2$, $U = -P(n - 1)an/(4h)$.

Supports reactions:

$X_A = Pa(n + 1)n/(4h)$, $Y_A = P(n + 1)$, $X_B = -X_A$.

Truss 5.10

A truss with a height of $3h$ and length $(2n+1)a$ consists of $4n+5$ rods (Fig. 163). The total length of the rods in the truss is

$$L_{sum} = (1 + 4n)a + (3n + 1)c + (n + 1)d + 3h,$$

where $c = \sqrt{a^2 + h^2}$, $d = \sqrt{a^2 + 9h^2}$.

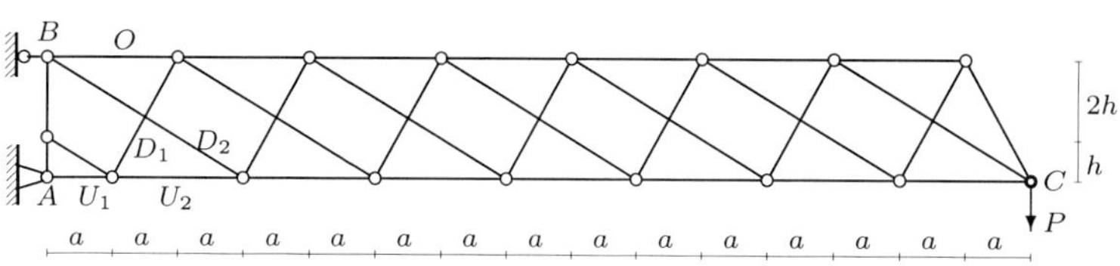

Fig. 163. Truss, $n = 7$

The deflection (vertical displacement of the node C) has the form

$$\Delta = P(C_1 a^3 + C_2 c^3 + C_3 d^3 + C_4 h^3)/(h^2 EF). \tag{5.11}$$

5.10.1. Concentrated force applied to the end of the console

Deflection. The coefficients in (5.11):

$C_1 = (16n^3 + 24n^2 + 26n - 6(-1)^n + 9)/27,$

$C_2 = 3(2n - (-1)^n + 1)/4,\ C_3 = (2n + 3(-1)^n + 1)/36,\ C_4 = 3.$

Forces:

$O = 2Pa(n-1)/(3h),\ D_1 = 0,\ D_2 = Pc/h,$

$U_1 = U_2 = -Pa(2n+1)/(3h).$

Supports reactions: $X_A = Pa(2n+1)/(3h),\ Y_A = P,\ X_B = -X_A.$

5.10.2. Load on the upper belt

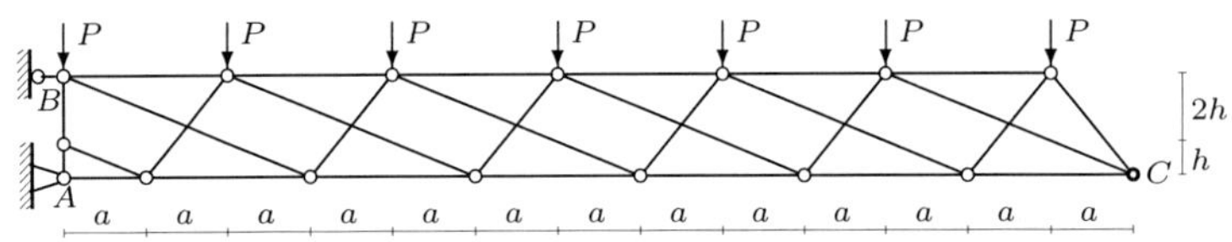

Fig. 164. Truss, $n = 6$

Deflection. The coefficients in (5.11):

$C_1 = n(2n^3 + 4n^2 + 7n - (-1)^n + 4)/9,$

$C_2 = 3(6n^2 + 2(3 - (-1)^n)n + 1 - (-1)^n)/16,$

$C_3 = (2n^2 + 2(1 + (-1)^n)n + (-1)^n - 1)/48,\ C_4 = 3(n+1).$

Forces:

$O = Pan(n-2)/(3h),\ D_1 = 0,\ D_2 = Pcn/h,$

$U_1 = U_2 = -Pa(n+1)n/(3h).$

Supports reactions:

$X_A = Pa(n+1)n/(3h),\ Y_A = P(n+1),\ X_B = -X_A.$

Chapter 6
Program

All the formulas given in the handbook are obtained using the symbolic mathematics program by induction on one, two or three independent parameters that specify the number of panels in parts of the structure. The Maple system was used here, although other software solutions, in particular the Mathematica package, can be used as well. Let us describe the algorithm for deriving formulas as applied to the case of one parameter. The author and his colleagues and students used this algorithm for solving various problems of statics for planar and spatial trusses. In [30, 31, 37] this method derived formulas for analytical estimates of truss vibration frequencies.

The algorithm of the method consists of two parts. First, in the analytical form, the deflection of a number of trusses with a successively increasing number of panels is calculated. Then a series of formulas obtained is generalized to an arbitrary number of panels. Calculation of the deflection of the truss with n_r rods and k rigid supporting rods is carried out according to the Maxwell-Mohr's formula

$$\Delta = \sum_{i=1}^{n_r-k} S_i^{(P)} S_i^{(1)} l_i/(EF), \tag{6.1}$$

where $S_i^{(P)}$ and l_i are the length and force in the ith rod from the load, $S_i^{(1)}$ is the force from the unit force applied to the central node, the displacement of which is calculated in the direction of the desired movement, E is the elastic modulus of the rods, F — cross-sectional area. Summation is carried out over all the rods of the truss, except for the k support, which is accepted to be undeformable. The solution is sought in an analytical form, therefore, the forces in the rods included in the Maxwell-Mohr's formula must also be sought in the form of formulas. For statically determined trusses, it is most convenient to find forces from the general system of equilibrium equations for all nodes (internal joints). The coefficients of the projection equations in this system are the guiding cosines of the forces. We give a block diagram of the program (Fig. 165):

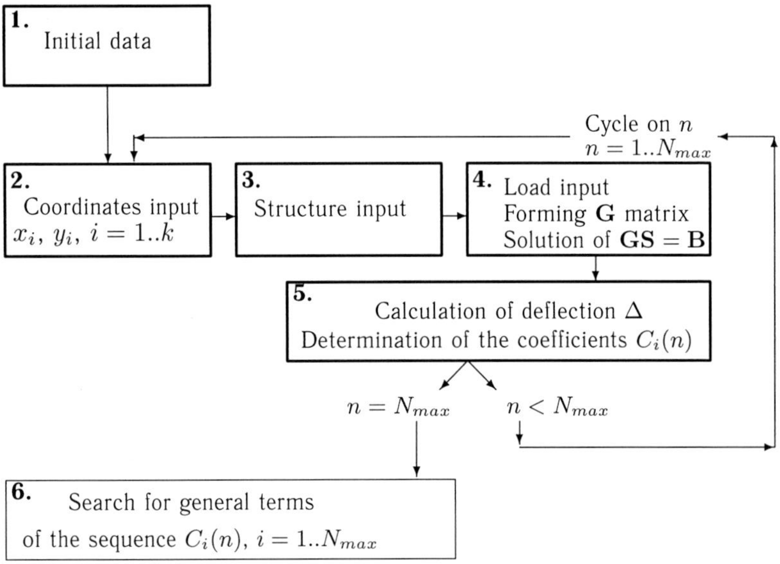

Fig. 165.

The forces in the rods of the truss are found from the solution of the system of equations of equilibrium of all nodes of the truss.

$$\mathbf{GS} = \mathbf{B}, \tag{6.2}$$

where $\mathbf{G}$ is the matrix of the directing cosines of the forces, S is the force vector in the truss rods, $\mathbf{B}$ is the load vector. The guiding cosines of the forces are determined based on the coordinates of the truss hinges (block **2**) and the lattice structure. The order of connecting the rods and hinges is specified by special vector rods with the numbers of the ends of the rods as components of the vectors (block **3**). For example, the entries $R_{1,8} = 9$, $R_{2,8} = 10$ mean that the rod with number 8 connects the hinges 9 and 10 (Fig. 166, 167). The choice of the direction of the vectors $\bar{R}$ does not affect the solution. To obtain the direction cosines, we first compute the projections of the conditional vectors $\bar{R}$ on the x and y axes

$$l_{1,i} = x_{R_{2,i}} - x_{R_{1,i}}, l_{2,i} = y_{R_{2,i}} - y_{R_{1,i}}$$

and the lengths of the rods

$$l_i = \sqrt{l_{1,i}^2 + l_{2,i}^2}.$$

The odd rows of the matrix of the system of equations include projections on the x axis, and the y rows on the y axis:

$$u = 2R_{i,2} - 2 + j,\ G_{u,i} = -l_{j,i}/l_i,$$

$$u = 2R_{i,1} - 2 + j,\ G_{u,i} = l_{j,i}/l_i,$$

$$u < n_r,\ j = 1, 2,\ i = 1, ..., n_r.$$

In block **5**, the formula (6.1) calculates the deflection and determines the coefficients in this formula for a given value of the number of panels n. Blocks **2**, **3**, **4**, and **5** are executed in a cycle of n. At the beginning (block **1**), the estimated number of N_{max} cycles is introduced into the program, which is enough for the program in block **6** to identify the general terms of the sequences of coefficients of the desired formula for deflection. The dimensions of the structure and the magnitude of the load are not specified; these quantities are uncertain task parameters. The determination of the general terms of sequences can be performed by discrete mathematics methods, or in some computer mathematics system. There are online solutions on the Internet for this operation, for example, based on WolframAlpha: https://allcalc.ru/node/974. If it is not possible to isolate the general terms, then in those cases when the determinant of the matrix $\mathbf{G}$ is independent of n, it suffices to increase the number of cycles N_{max}.

We give an example of the derivation of the formula for the dependence of the deflection of the cantilever truss, 5.8, p. 150 with a cross-shaped lattice of the number of panels under the action of a uniform load. In block **1**, you need to set the number of cycles. For this truss, $N_{max} = 12$ is sufficient [1].

[1] in some cases, the required number of cycles reaches 40.

Setting the coordinates of the hinges (block **2**, Fig. 166) has the form:

$$x_i = x_{i+n+1} = (i-1)a, y_i = 0,$$
$$y_{i+n+1} = h, i = 1, ..., n+1.$$

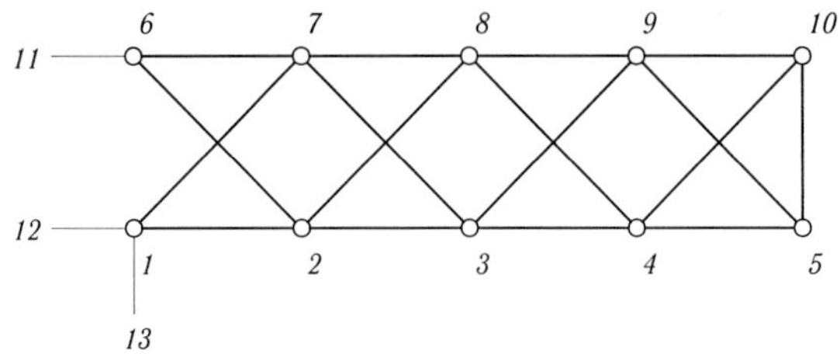

Fig. 166. Node Numbering, $n = 4$

The assignment of the numbers of the rods of the lower and upper zones (structure entry, block **2**, Fig. 167) has the form

$$R_i = [i, i+1],$$
$$R_{i+n} = [i+n+1, i+n+2],\ \ i = 1, ..., n.$$

The truss lattice is defined by the following vectors

$$R_{i+2n} = [i+1, i+n+1],$$
$$R_{i+3n} = [i, i+n+2],\ \ i = 1, ..., n,$$
$$R_{4n+1} = [2n+2, n+1].$$

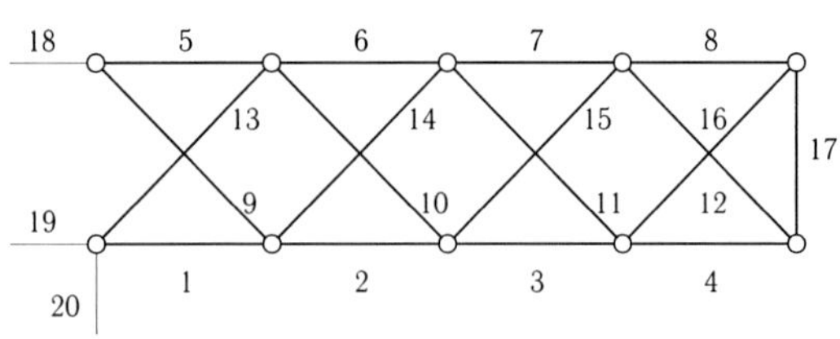

Fig. 167. Rods Numbering, $n = 4$

Table 1 shows the truss schemes and formulas for deflection obtained by sequential calculation of trusses with $n = 1, 2, 3$ panels. In fact, 12 trusses are required to identify general term of the sequences of coefficients. The deflection value at $n = 12$ (Fig. 168) has the form:

$$\Delta = 40P(146a^3 + 3c^3)/(EFh^2).$$

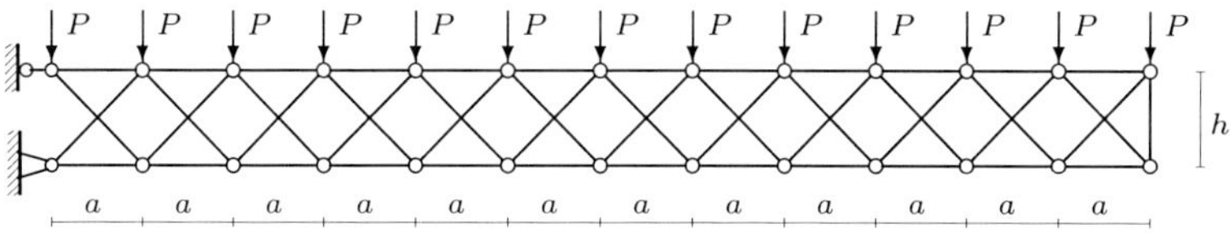

Fig. 168. Truss, $n = 12$

Table 1

n	Scheme	Δ
1	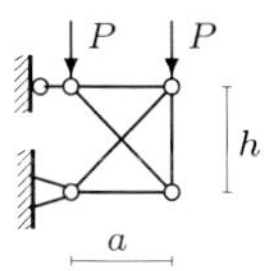	$\Delta = P(2a^3 + 2c^3 + h^3)/(EFh^2)$,
2	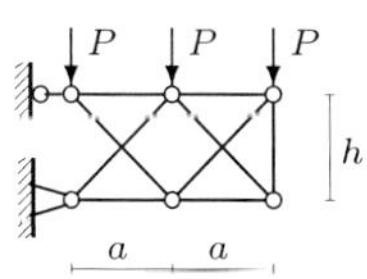	$\Delta = 5P(2a^3 + c^3)/(EFh^2)$,
3	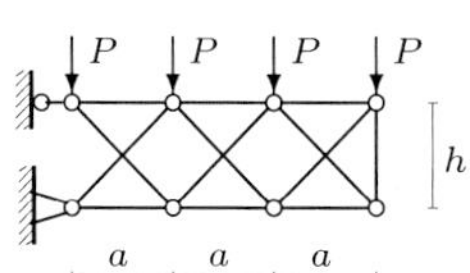	$\Delta = 2P(18a^3 + 5c^3 + h^3)/(EFh^2)$
..	...	Δ

General view of the expression for deflection:

$$\Delta = P(C_1a^3 + C_2c^3 + C_3h^3)/(EFh^2).$$

We write down the sequence of coefficients C_1:

2, 10, 36, 96, 214, 418, 744, 1232, 1930, 2890, 4172, 5840, ...

The Maple system has an operator rgf_findrecur, which gives the following recurrence equation for these coefficients:

$$C_{1,n} = 4C_{1,n-1} - 5C_{1,n-2} + 5C_{1,n-4} - 4C_{1,n-5} + C_{1,n-6}.$$

Its solution (5.9), p. 151 gives the rsolve operator of the same system. Similar operators are in the Mathematica computer mathematics system.

Similarly, the coefficients C_2 and C_3 of the desired formula are obtained. In cases where the truss has a second regularity parameter m, for example, the number of vertical panels in the frame (truss 2.1, p. 41), it is necessary to perform an additional m cycle. The systems with three parameters (truss 2.3, p. 45) are more complex.

Trusses Schemes

No.	Scheme	p.
	Arches	
1.1	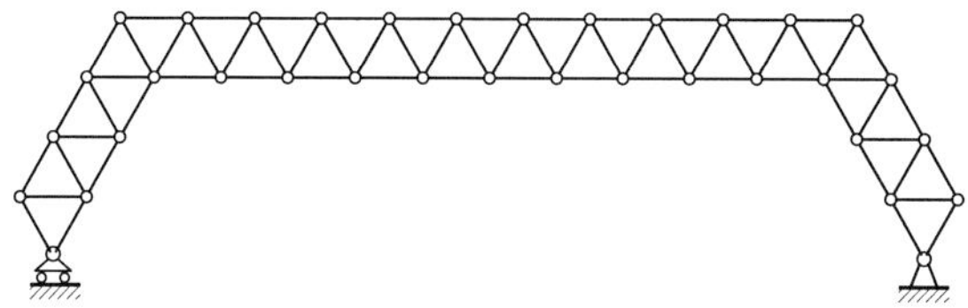	p. 4
1.2	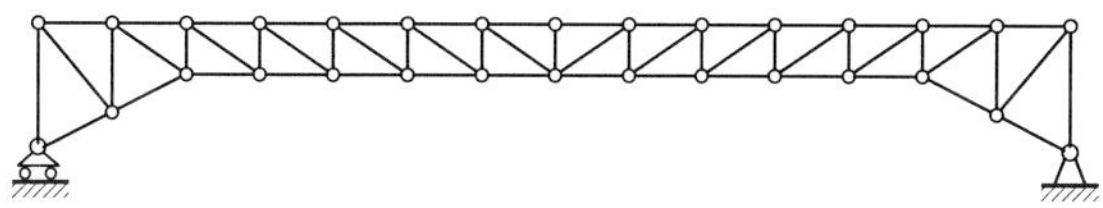	p. 5
1.3	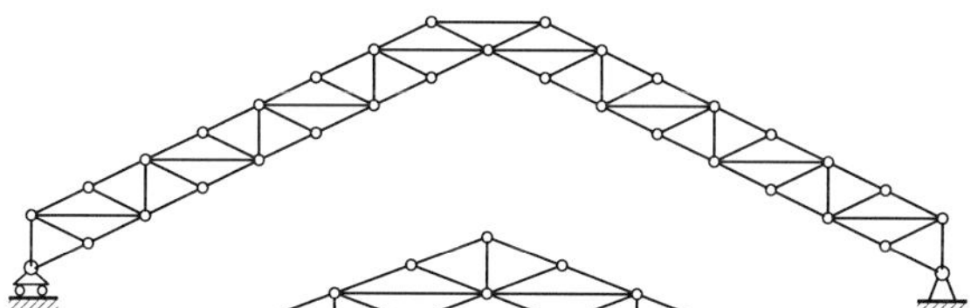	p. 7
1.4	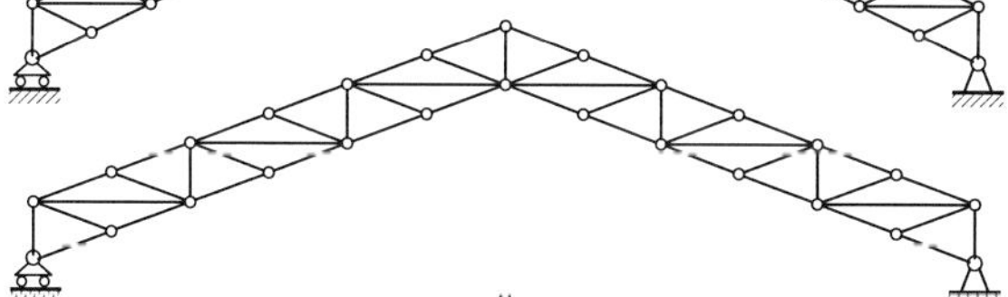	p. 9
1.5	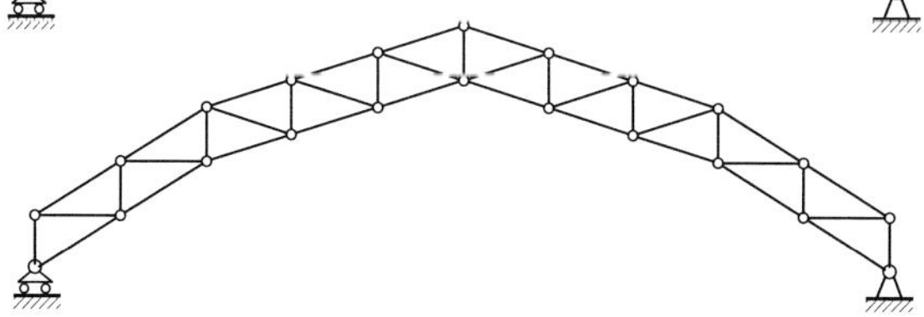	p. 11
1.6	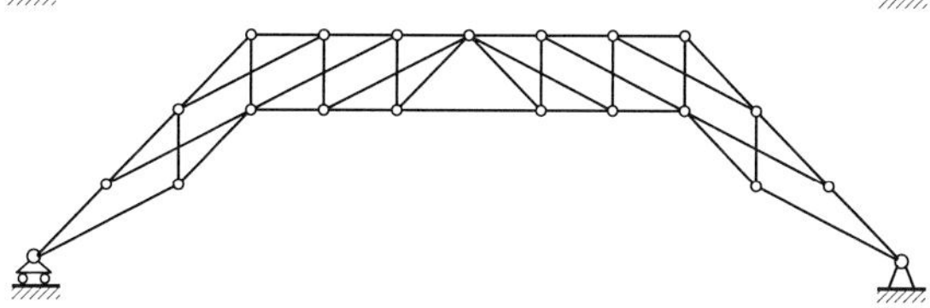	p. 13

No.	Scheme	p.
1.7	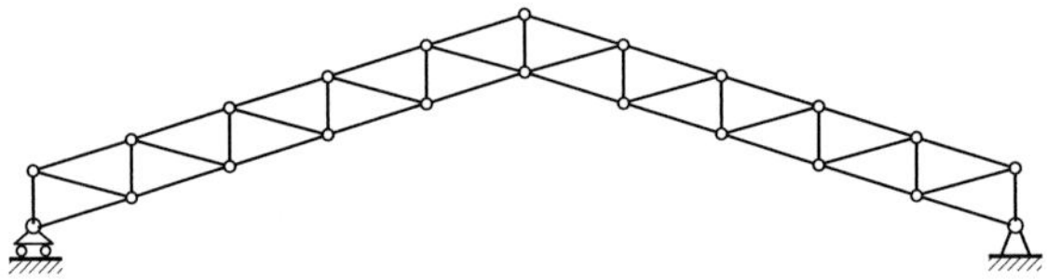	p. 17
1.8	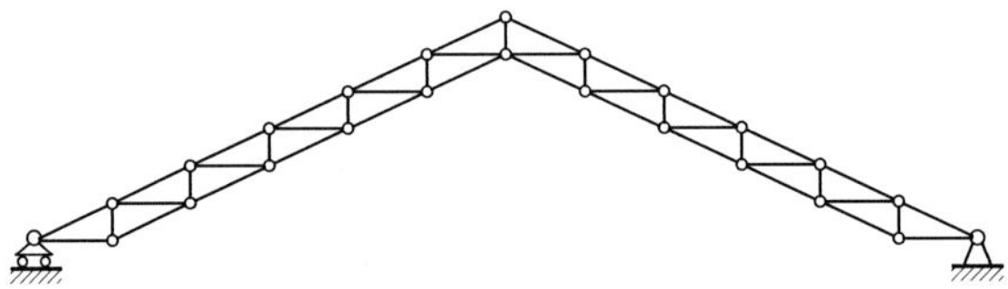	p. 19
1.9	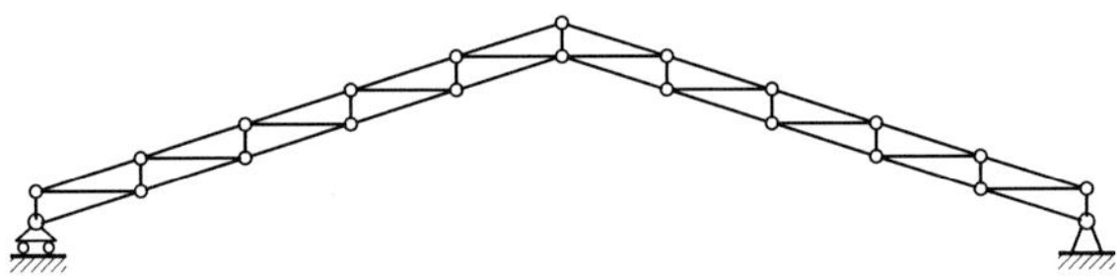	p. 21
1.10	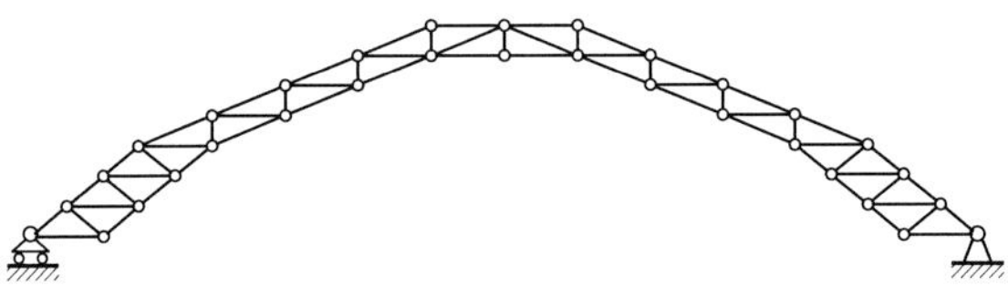	p. 22
1.11	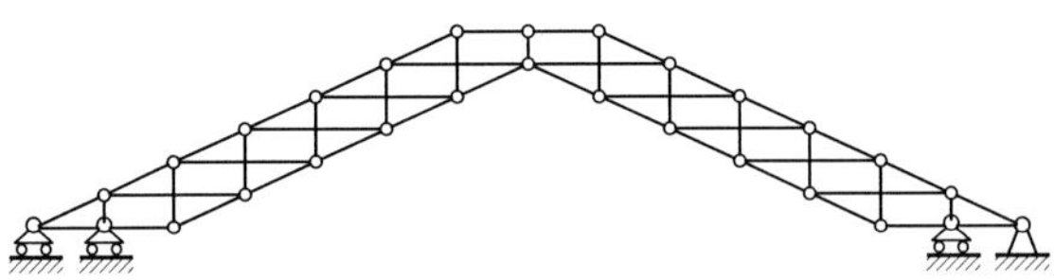	p. 25
1.12	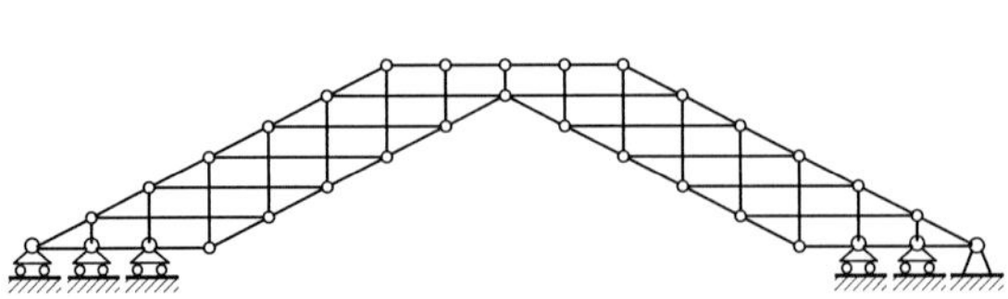	p. 27

No.	Scheme	p.
1.13	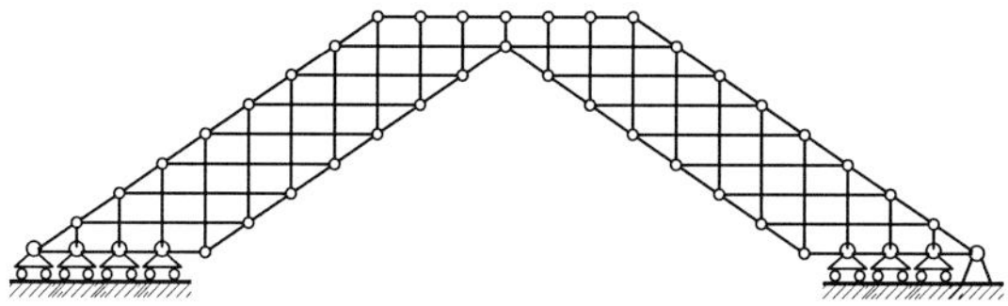	p. 29
1.14	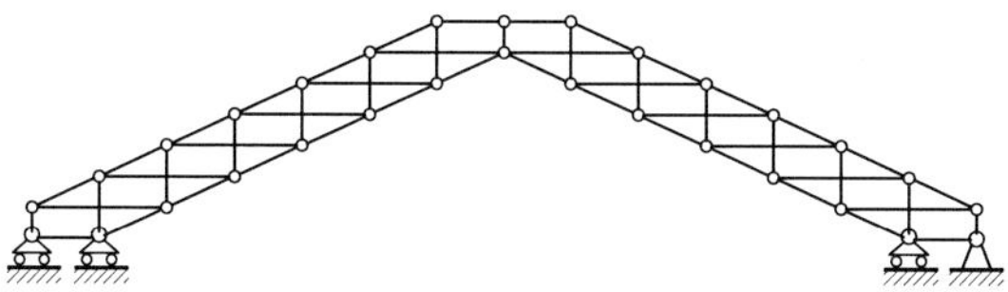	p. 31
1.15	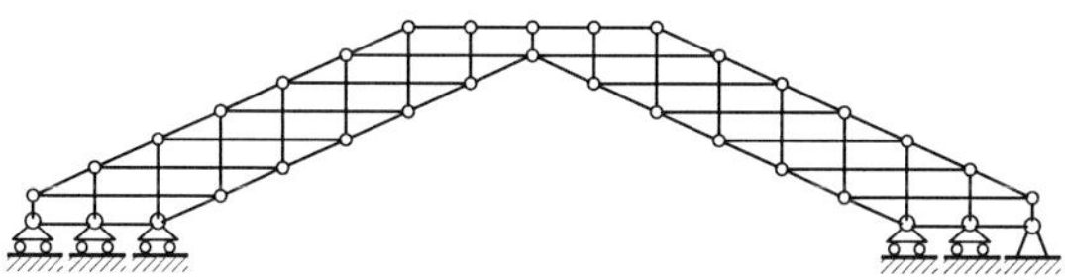	p. 33
1.16	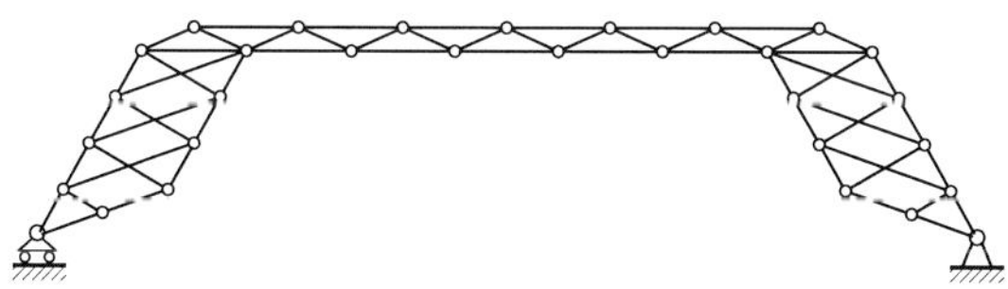	p. 35
1.17	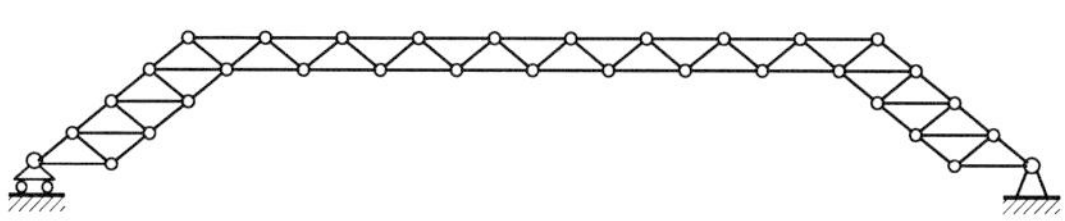	p. 38
	Frames	
2.1	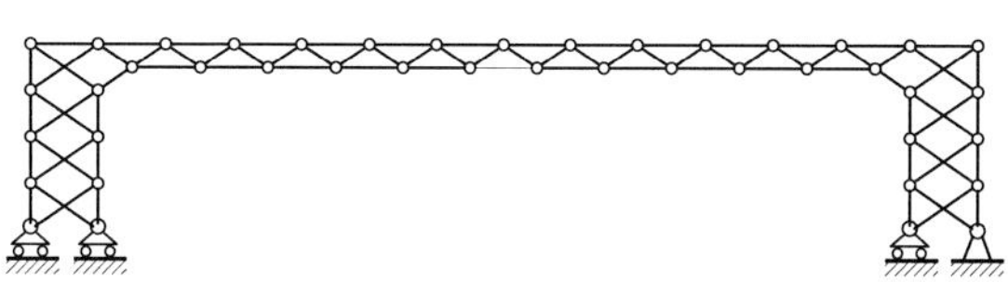	p. 41

No.	Scheme	p.
2.2	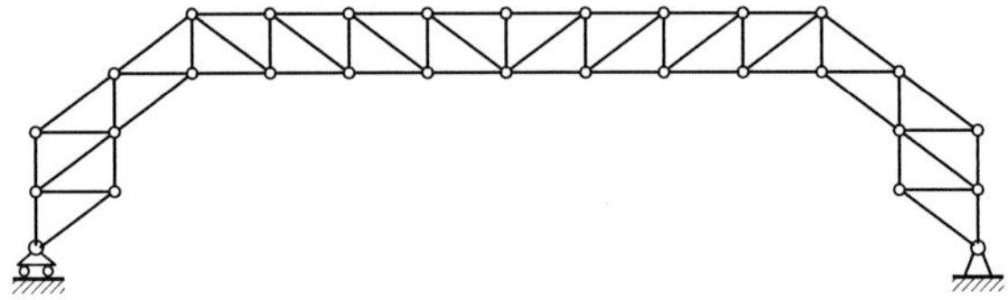	p. 43
2.3	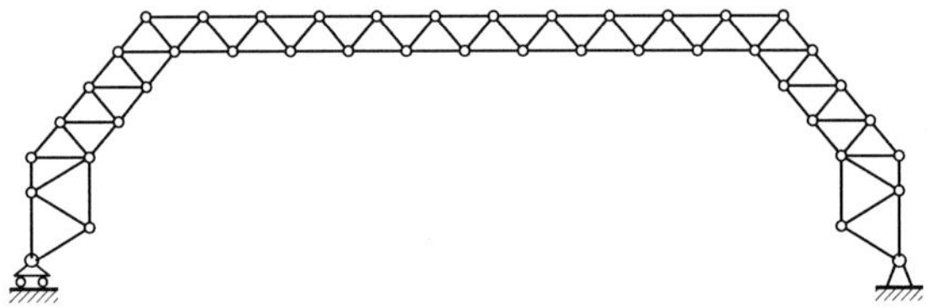	p. 45
2.4	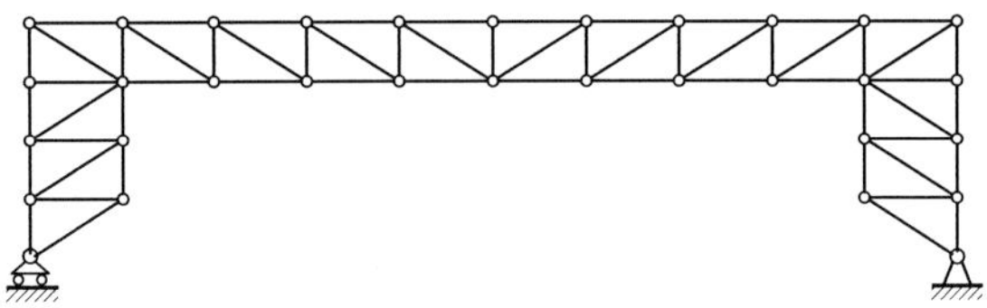	p. 47
2.5	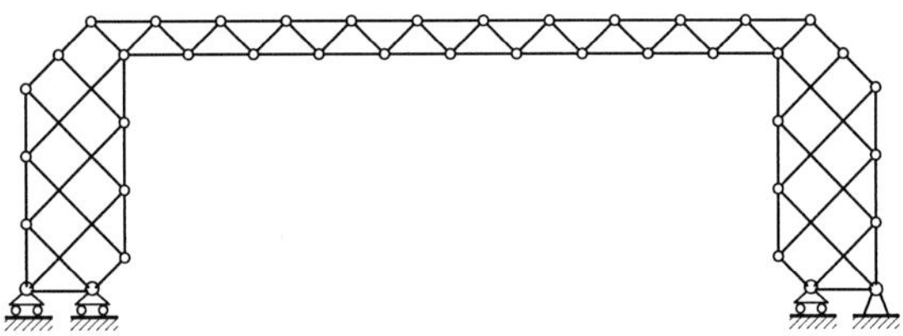	p. 49
2.6	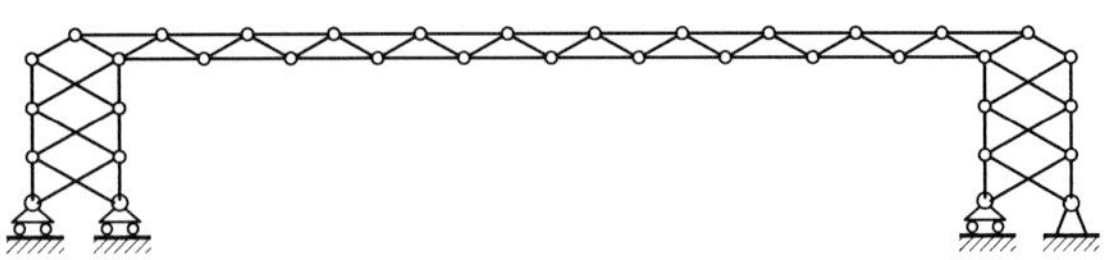	p. 51
2.7	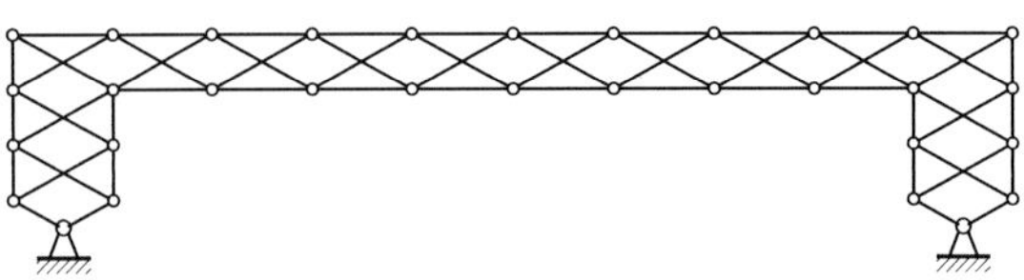	p. 53

No.	Scheme	p.
2.8	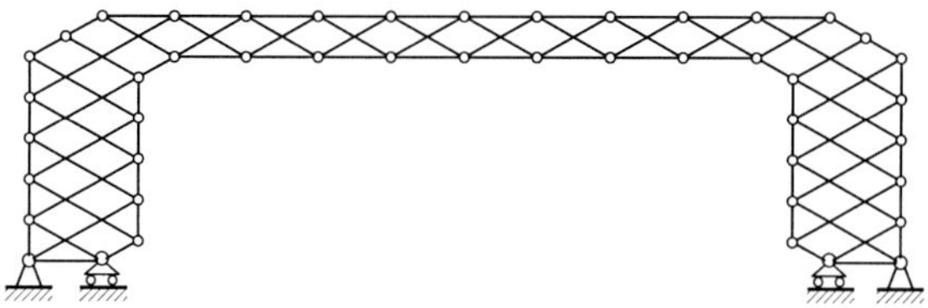	p. 56
2.9	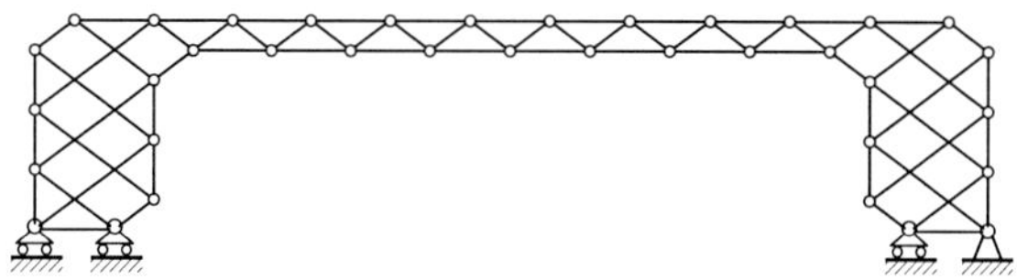	p. 59
2.10	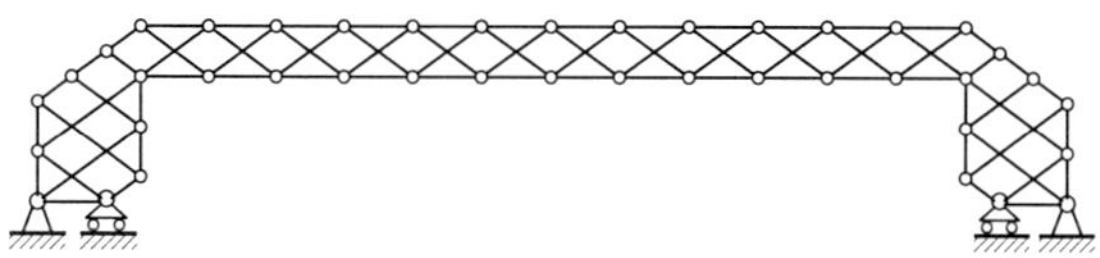	p. 62
2.11	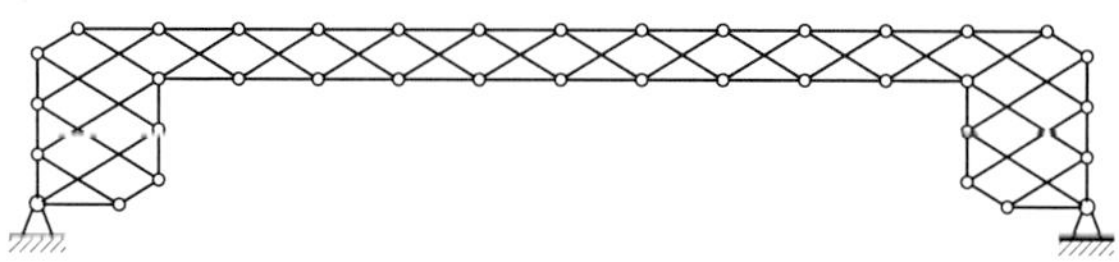	p. 63
2.12	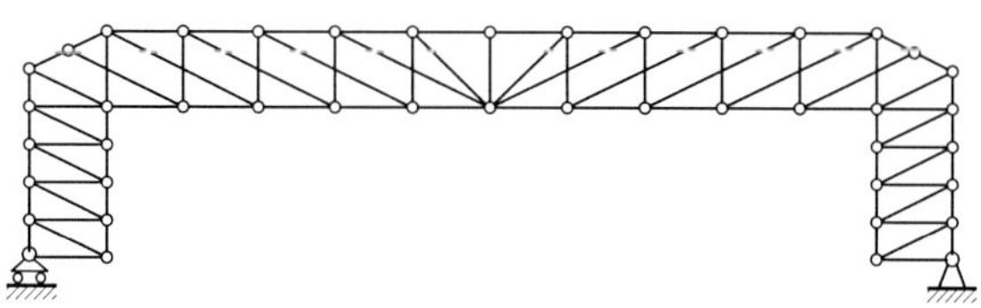	p. 66
2.13	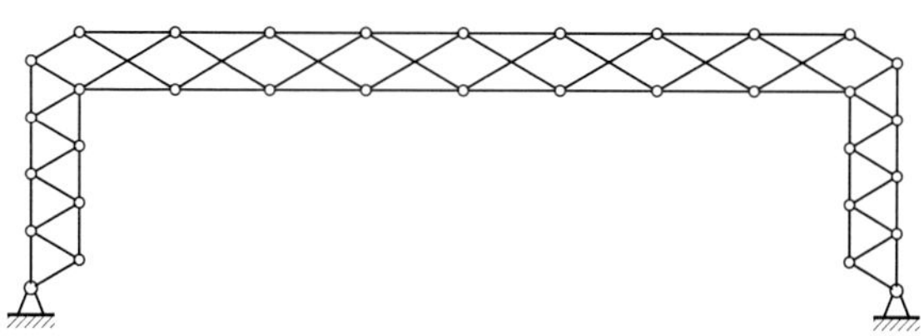	p. 68

No. Scheme p.

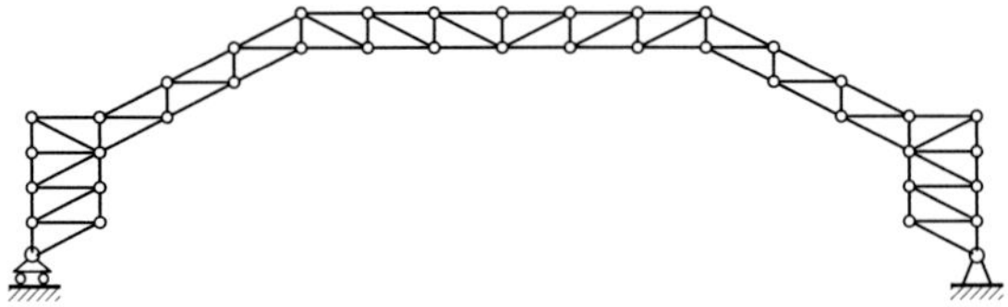

2.14 p. 70

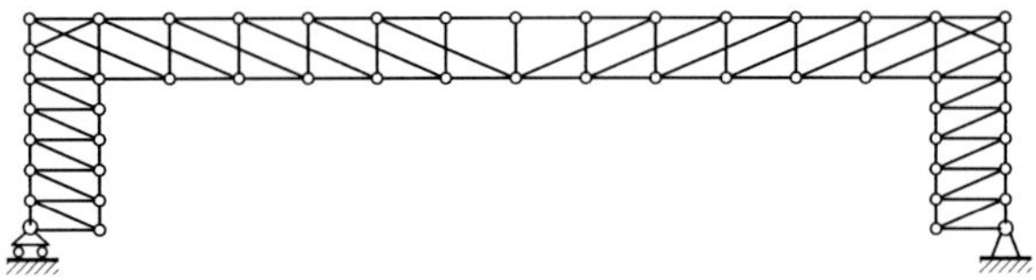

2.15 p. 72

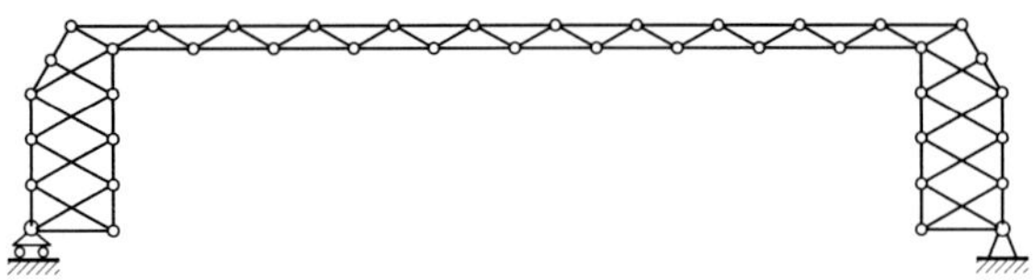

2.16 p. 74

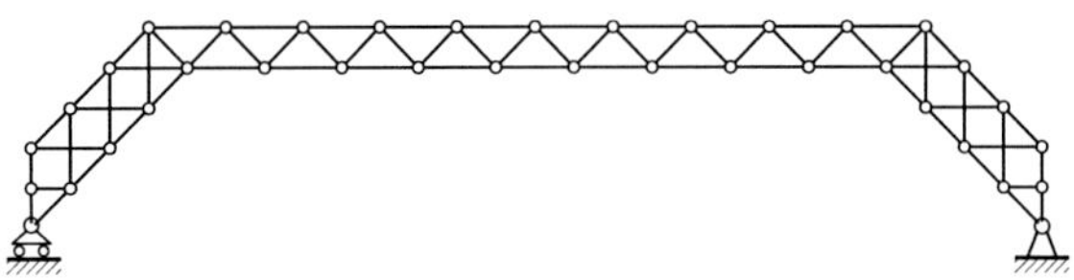

2.17 p. 76

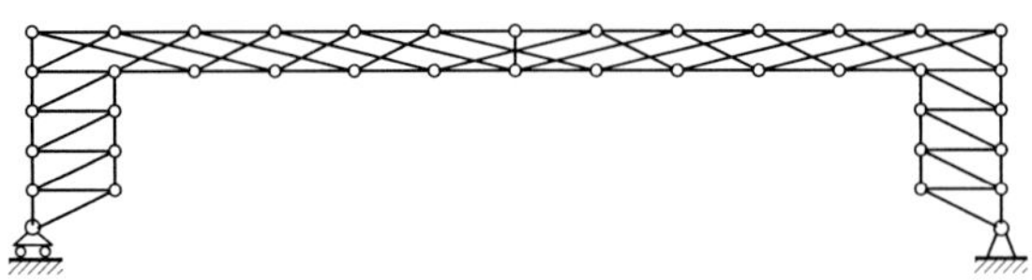

2.18 p. 78

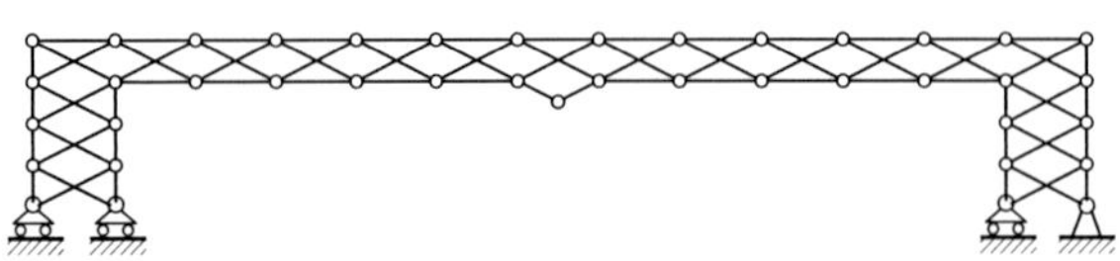

2.19 p. 80

No. Scheme p.

Composite Frames and Arches

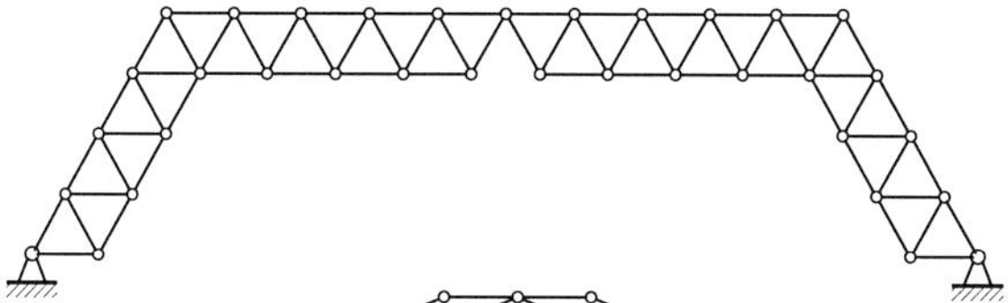

3.1 p. 84

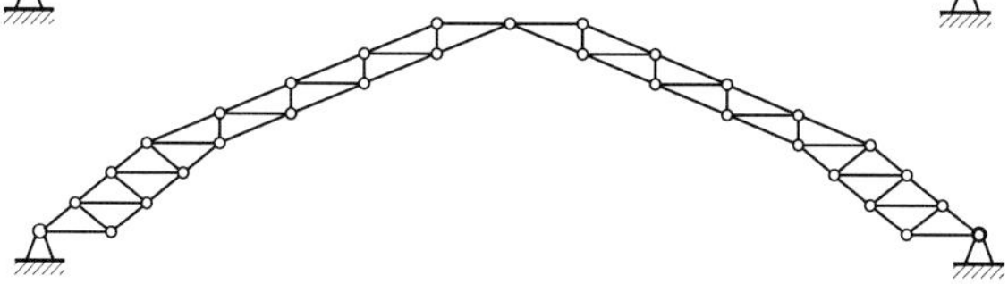

3.2 p. 85

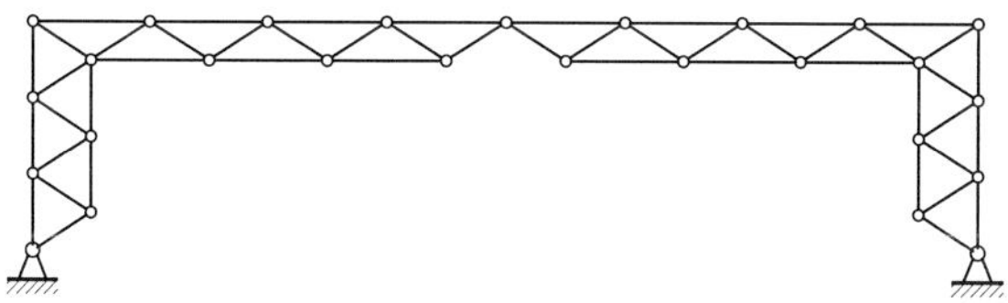

3.3 p. 87

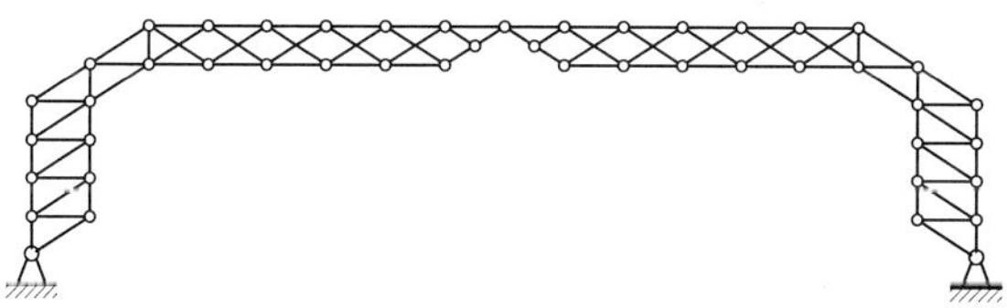

3.4 p. 89

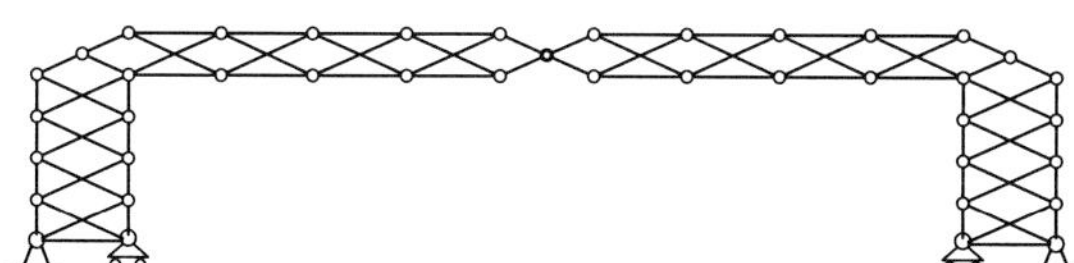

3.5 p. 91

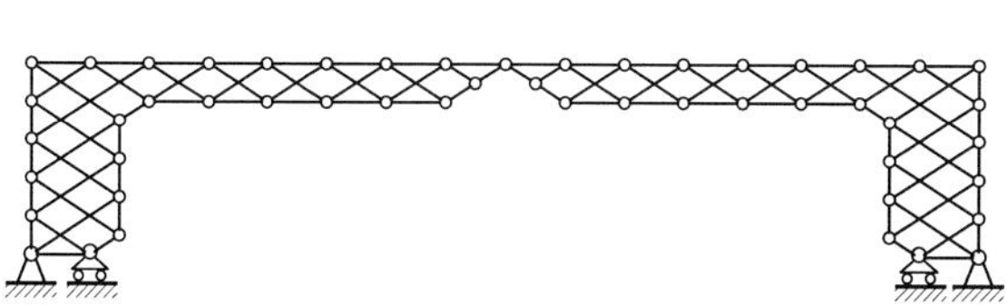

3.6 p. 93

No.	Scheme	p.
3.7	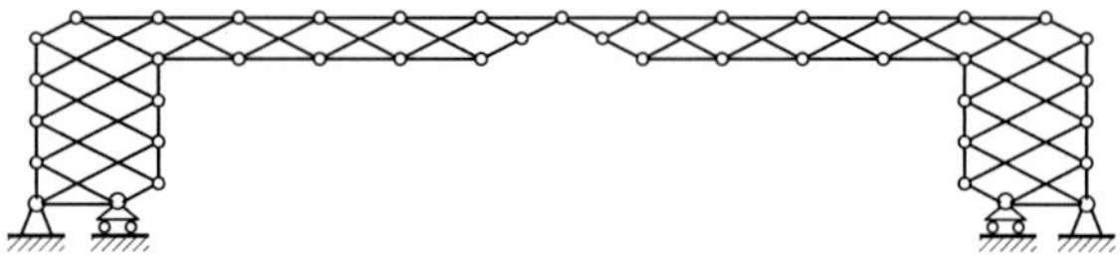	p. 95
3.8	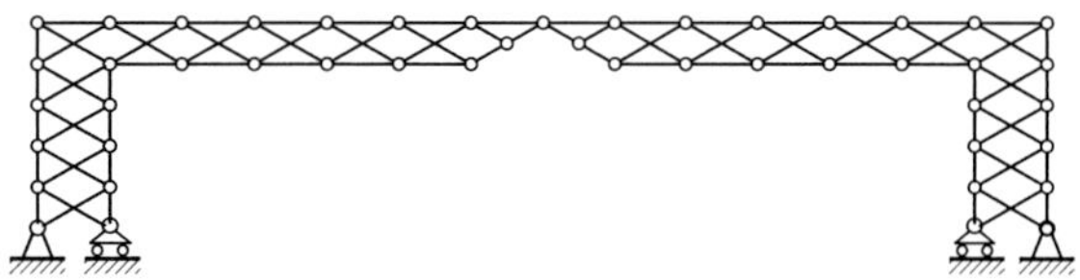	p. 97
3.9	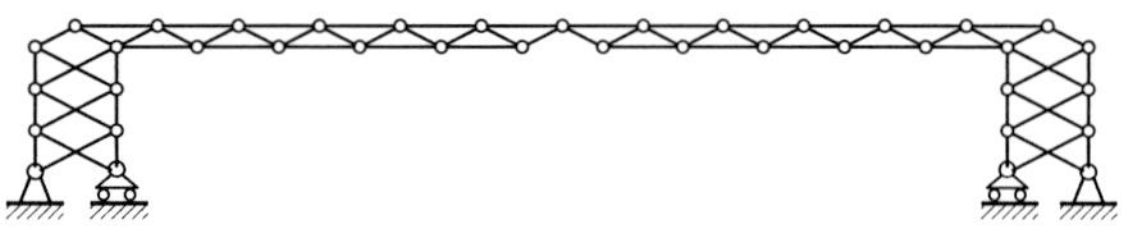	p. 99
3.10	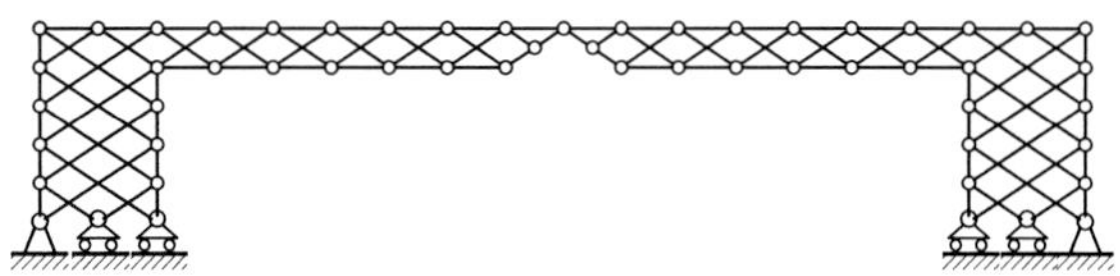	p. 102
3.11	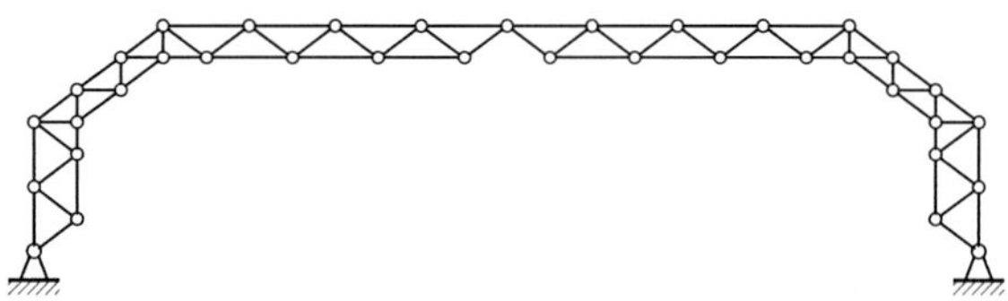	p. 105
3.12	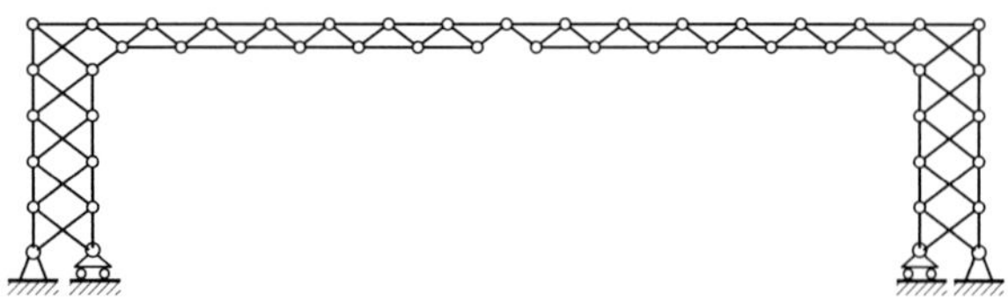	p. 107

No. Scheme p.

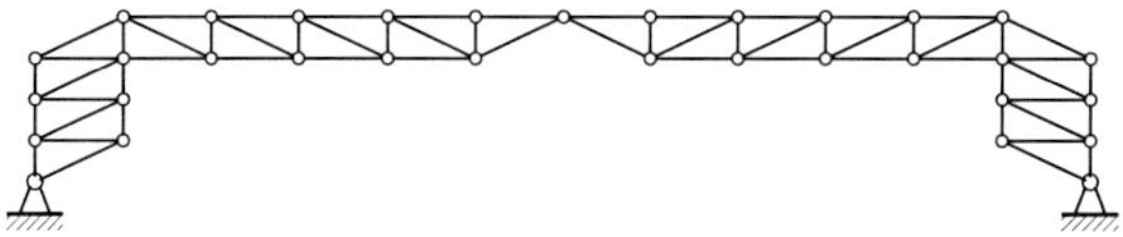

3.13 p. 109

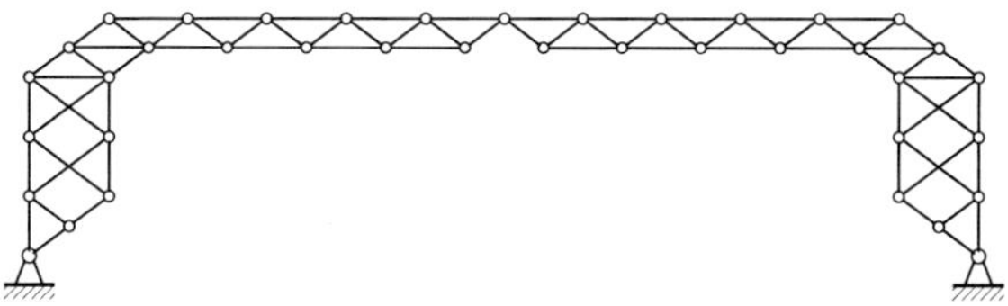

3.14 p. 111

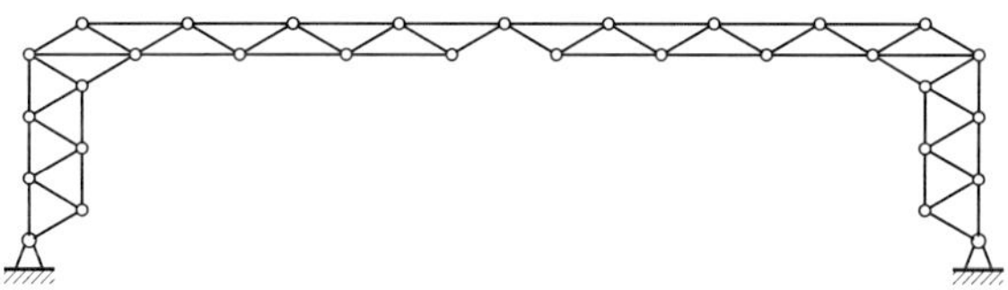

3.15 p. 113

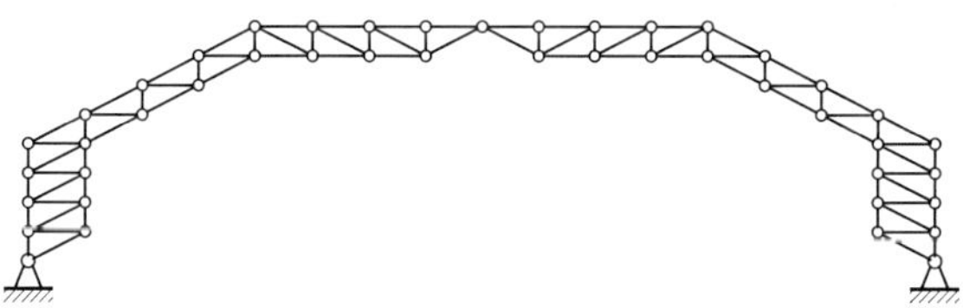

3.16 p. 115

Multi-span Frames and Brackets

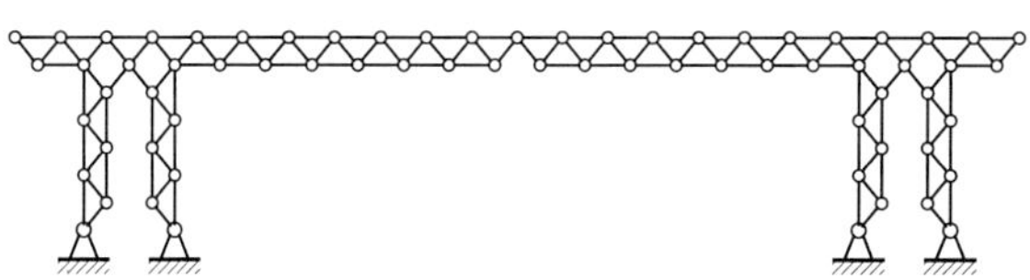

4.1 p. 117

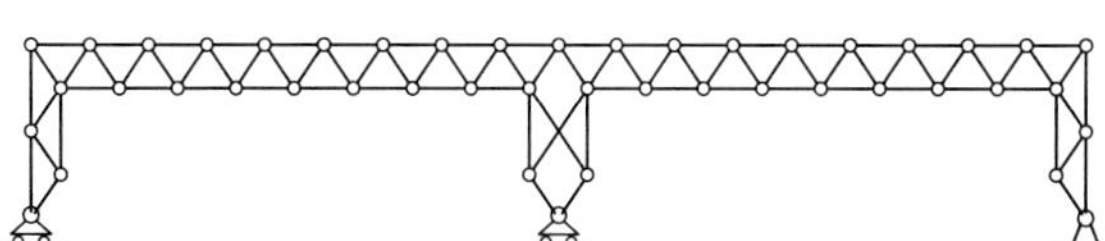

4.2 p. 119

No.	Scheme	p.
4.3	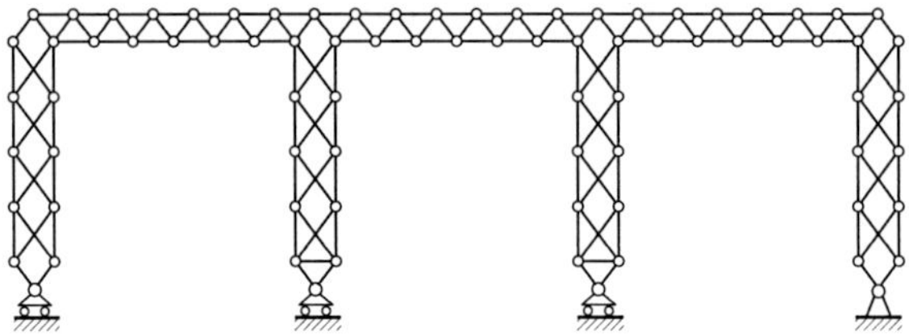	p. 122
4.4	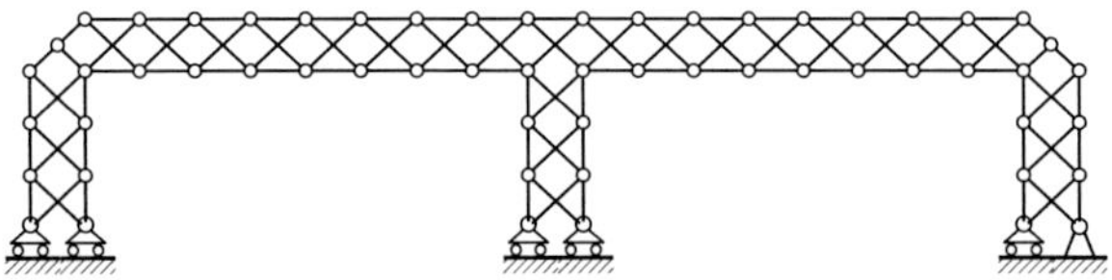	p. 124
4.5	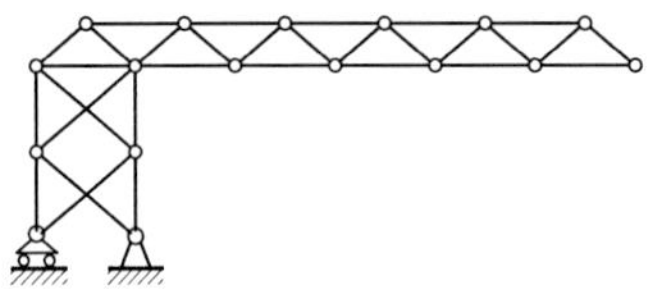	p. 126
4.6	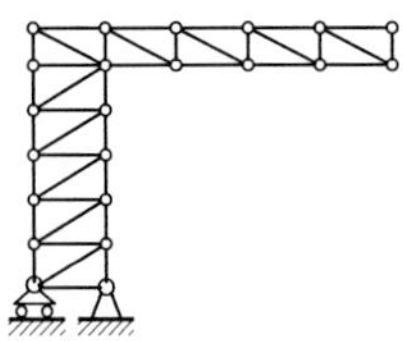	p. 128
4.7	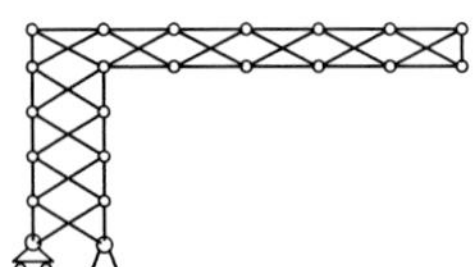	p. 129
4.8	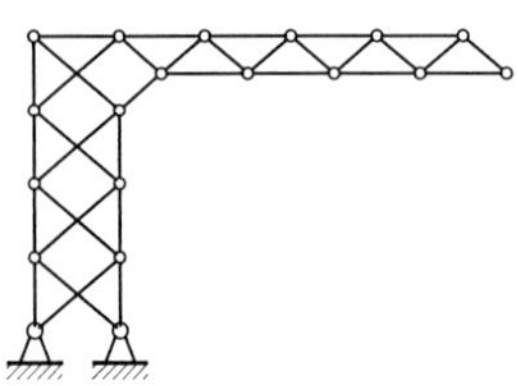	p. 131

No.	Scheme	p.
4.9	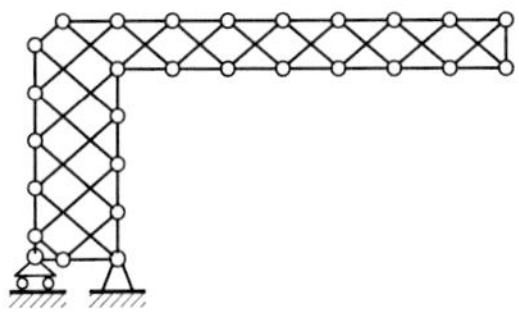	p. 133
4.10	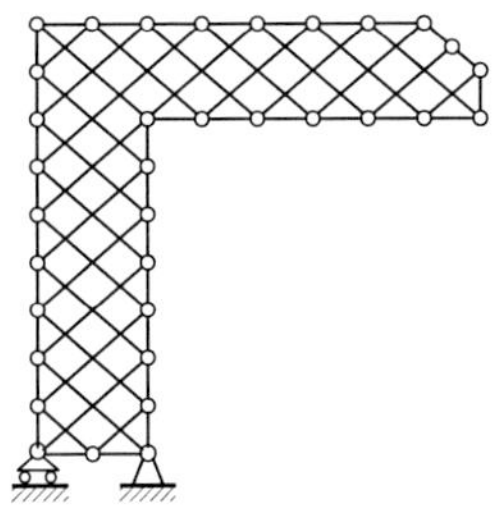	p. 135
	Cantilever trusses	
5.1	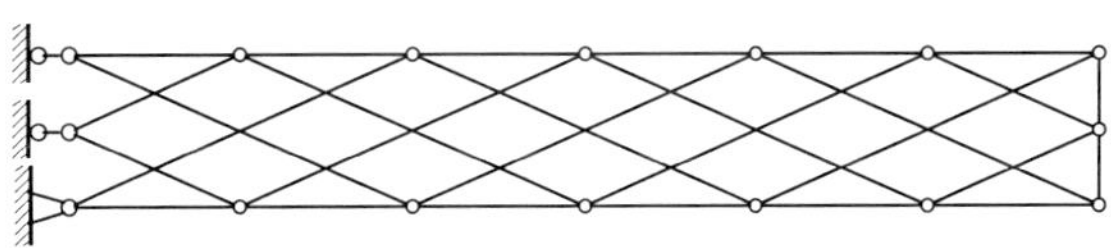	p. 139
5.2	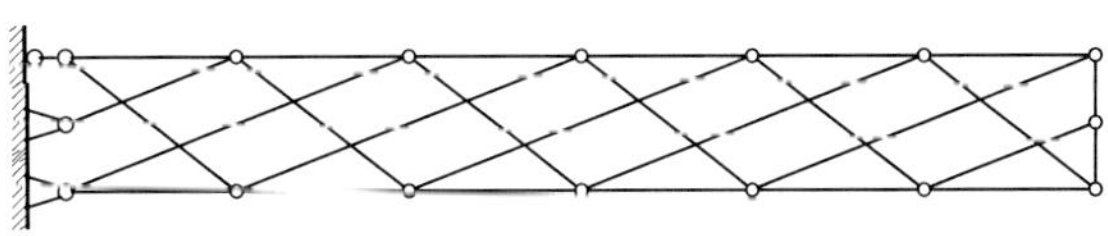	p. 141
5.3	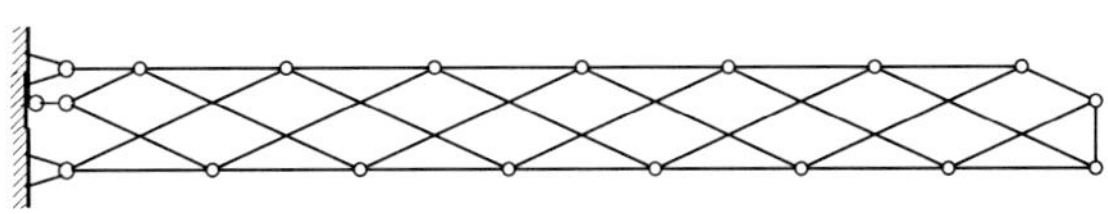	p. 143
5.4	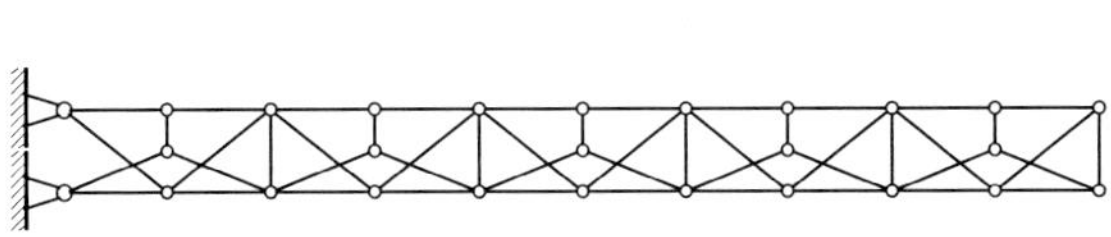	p. 145

No.	Scheme	p.
5.5	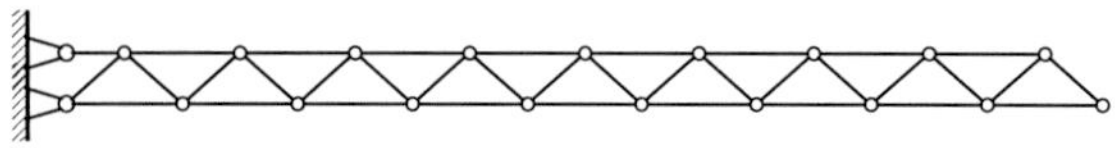	p. 146
5.6	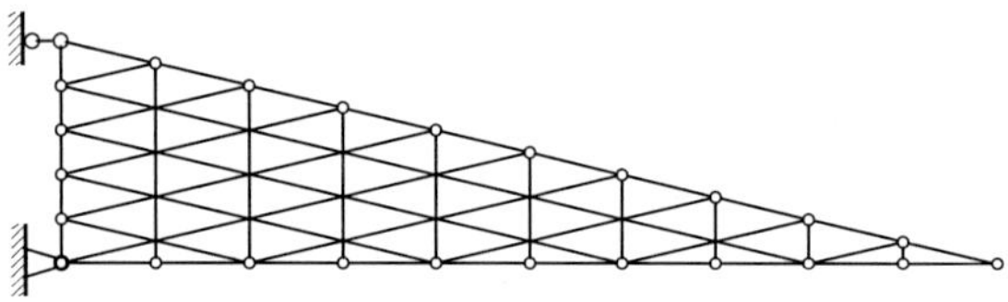	p. 148
5.7	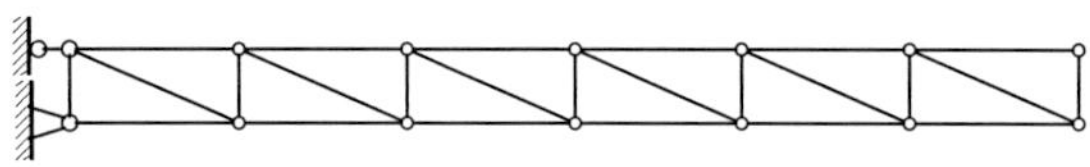	p. 149
5.8	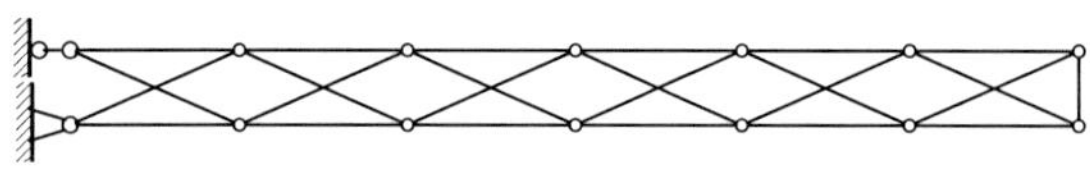	p. 150
5.9	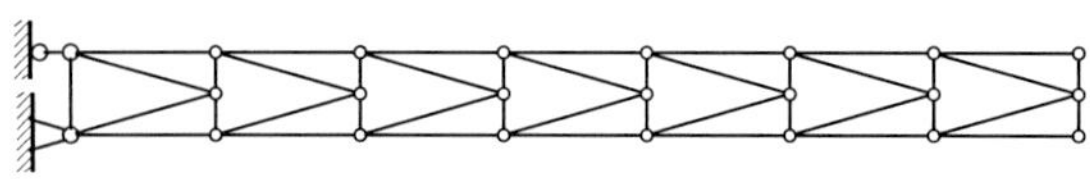	p. 152
5.10	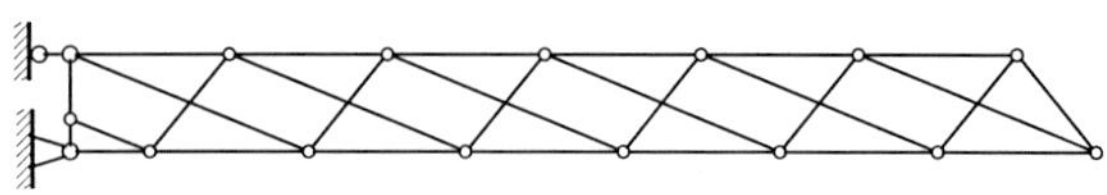	p. 153

References

1. *Al-Shahrabi A.M., Kirsanov M.N.* Line of influence of the deflection for cantilever truss. Bulletin of Scientific Conferences. 2016. No. 2-1(6). Pp. 6-7. http://ucom.ru/doc/cn. 2016.02.01.pdf
2. *Arutyunyan V.B.* Calculation of the deflection of a statically indeterminate beam truss. Postulat 2018. 6(32).
3. *Arutyunyan V.B.* Double induction for deriving a formula for deflecting a frame truss with an arbitrary number of panels. Postulat. 2018. 7(33).
4. *Belyankin N.A., Boyko A.Y., Kirsanov M.N.* Inductive derivation of rod cladding structures deflection in the system Maple. Science Almanac. 2017. No. 4-3(30). Pp. 194-197. doi://10.17117/na. 2017.04.03.194
5. *Belyankin N.A., Boyko A.Y.* Analysis of the deflection of the flat statically determinate girder. Science Almanac. 2017. No. 2-3(28). Pp. 246-249.
6. *Bolotina T.D.* The deflection of the flat arch truss with a triangular lattice depending on the number of panels. Bulletin of Scientific Conferences. 2016. No. 4-3(8). Pp.7-8.
7. *Buka-Vaivade K., Kirsanov M.N., Serdjuks D.O.* Calculation of deformations of a cantilever-frame planar truss model with an arbitrary number of panels. Vestnik MGSU. Monthly Journal on Construction and Architecture. 2020; 15(4). Pp. 510-517. doi: 10.22227/1997-0935.2020.4.510-517
8. *Chinchenko A. M.* Analytical calculation of the size of the bend of rectangular shape frame truss with a random number of panels in the Maple system. Mechanization and automation of construction [Electronic resource]: collection of articles. Samara: Samar state technical University. Un-t, 2019. Pp. 54-58.
9. *Domanov E.V.* The dependence of the deflection of the cantilever truss on the number of panels obtained in the system Maple. Structural mechanics and structures. 2018. 2(17). Pp. 80-86.
10. *Domanov E.V.* The formula for the dependence of the deflection of an externally statically indeterminate truss on the number of panels. Postulat. 2018. 6(32).
11. *Domanov E.V.* An analytical solution of the problem of the externally statically indeterminate truss deflection with an arbitrary number of panels. Postulat. 2017. 7(21).

12. *Domanov E.V.* Finding the dependence of the deflection of a flat beam-cantilevered truss on the number of panels in the Maple system. Postulat. 2018. No. 12 (38).

13. *Domanov E.V., Kirsanov M.N.* Analytical calculation of the deflection of the truss with cross-bar lattice. Science Almanac. 2016. No. 11-2(25). Pp. 312-314. doi: 10.17117/na. 2016.11.02.312

14. *Doneva A., Torquatob S.* Energy-efficient actuation in infinite lattice structures.Journal of the Mechanics and Physics of Solids. 2003. Vol.51. No. 8. Pp. 1459–1475.

15. *Dong Xiaomeng, Kirsanov M.N.* The dependence of the deflection of the truss from the position of the load for an arbitrary number of panels. Science Almanac. 2016. No. 1-4(5). Pp. 6-7.

16. *Hutchinson R.G., Fleck N.A.* Microarchitectured cellular solids – the hunt for statically determinate periodic trusses. ZAMM Z.Angew.Math.Mech. 2005. 85. No. 9. Pp. 607 – 617.

17. *Hutchinson R.G., Fleck N.A.* The structural performance of the periodic truss. Journal of the Mechanics and Physics of Solids. 2006. Vol. 54. Issue 4. Pp.756-782.

18. *Jiang H., Kirsanov M.N.* An analytical expression for the influence line of the truss. Bulletin of Scientific Conferences. 2016. No. 1-5(5). Pp. 10-11.

19. *Ilin I.A., Kirsanov M.N.* Analysis of the deflection of externally statically indeterminate truss. Science Almanac. 2017. No. 2-3(28). Pp. 260-263. doi: 10.17117/na.2017.02.03.260

20. *Ilin I.A., Kirsanov M.N.* The deflection and displacement of the bearings of the truss with rhombic lattice. Science Almanac. 2016. No. 12-2(26). Pp. 216-219. doi: 10.17117/na.2016.12.02.216

21. *Ilyushin A.S.* The formula for calculating the deflection of a compound externally statically indeterminate frame. Structural mechanics and strength of materials. 2019. Vol. 3. No. 22. Pp. 29-38.

22. *Kaveh A, Shahrouzi M., Farshi B., Alinia-Ziazi A.* Sizing optimization of truss structures by method of centers and force formulation. International Journal of Solids and Structures. 2010. Vol. 47. No. 18-19. Pp. 2508–2524.

23. *Kitaev S.S.* Derivation of the formula for the deflection of a cantilevered truss with a rectangular diagonal grid in the computer mathematics system Maple. Postulat. 2018. No. 5-1(31).

24. *Kirsanov M.N.* Analytical dependence of the deflection of the spatial truss on the number of panels. Magazine of Civil Engineering. 2020. 96(4). Pp. 110–117. doi: 10.18720/MCE.96.9

25. *Kirsanov M.N., Serdjuks D., Buka-Vaivade K.* Analytical Expression of the Dependence of the Multi-lattice Truss Deflection on the Number of Panels. Construction of Unique Buildings and Structures. 2020. Vol. 90. Article No 9003. doi: 10.18720/CUBS.90.3
26. *Kirsanov M.N.* Analytical Solution of a Spacer Beam Truss Deflection with an Arbitrary Number of Panels. Construction of Unique Buildings and Structures. 2020. Vol. 88. Article No 8802. doi: 10.18720/CUBS.88.2.
27. *Kirsanov M., Serdjuks D., Buka-Vaivade K.* Analytical Dependence of Deflection of the Lattice Truss on the Number of Panels. Springer Nature Switzerland AG 2020 B. Anatolijs et al. (eds.), Proceedings of EECE 2019, Lecture Notes in Civil Engineering 70, Pp. 25-35. https://doi.org/10.1007/978-3-030-42351-3_3
28. *Kirsanov M.N.* Planar Trusses: Schemes and Formulas. Cambridge Scholars Publishing. 2019.
29. *Kirsanov M.N., Buka-Vaivade K.* Analytical calculation of the deflection of the rod frame with an arbitrary number of panels. Structural mechanics and strength of materials. 2019. Vol. 3. No. 22. Pp. 21-28.
30. *Kirsanov M.N., Buka-Vaivade K.* Analytical expressions of frequencies of small oscillations of a beam truss with an arbitrary number of panels. Structural mechanics and strength of materials. 2019. Vol. 4. No. 23. Pp. 7-14.
31. *Kirsanov M.N.* Lower estimate of the fundamental frequency of natural oscillations of a truss with an arbitrary number of panels. Proceedings of Moscow State University of Civil Engineering. 2019. 14.7. 844-851. doi: 10.22227/1997-0935.2019.7.844-851
32. *Kirsanov M.N.* Calculation of G-shaped truss of Road Signs and Equipment. Russian Journal of Building. Construction and Architecture. 2019. No. 2(42). Pp. 62-71. doi 10.25987/VSTU.2019.42.2.007
33. *Kirsanov M.N.* Parallelogram Mechanism with Any Number of Sections. Russ. Engin. Res. 2018. Vol.38. No.4. Pp. 268–271. https://doi.org/10.3103/S1068798X18040135
34. *Kirsanov M.N.* Analytical calculation of the frame with an arbitrary number of panels. Magazine of Civil Engineering. 2018. 82(6). Pp. 127–135. doi: 10.18720/MCE.82.12
35. *Kirsanov M.N.* Installation diagram of the lattice truss with an arbitrary number of panels. Magazine of Civil Engineering. 2018. 81(5). Pp. 175–183. doi: 10.18720/MCE.81.17
36. *Kirsanov M., Tinkov D., Boiko O.* Analytical calculation of the combined suspension truss. MATEC Web of Conferences, Vol. 265. 0525 (2019) https://doi.org/10.1051/matecconf/201926505025

37. *Kirsanov M.N., Tinkov D.V.* Analytical solution of the frequency of the load oscillation at an arbitrary girder node in the system Maple. Construction: Science and Education. 2018. 8(4):3. URL: http://nso-journal.ru. doi: 10.22227/2305-5502.2018.4.3.

38. *Kirsanov M.N., Tinkov D.V.* Analytical calculation of the deflection of the lattice truss. MATEC Web of Conferences. 2018. Vol. 193.03015. doi.org/10.1051/matecconf/201819303015

39. *Kirsanov M.N., Komerzan E., Sviridenko O.* Inductive analysis of the deflection of a beam truss allowing kinematic variation. MATEC Web of Conferences. 2018. Vol. 239, 01012.https://doi.org/10.1051/matecconf/201823901012

40. *Kirsanov M.N.* A Precise Solution of the Task of a Bend in a Lattice Girder with a Random Number of Panels. Russian Journal of Building Construction and Architecture. 2018. No. 1(37). Pp. 92-99.

41. *Kirsanov M.N.* The deflection of spatial coatings with periodic structure. Magazine of Civil Engineering. 2017. No. 8. Pp. 58-66. doi: 10.18720/MCE.76.6

42. *Kirsanov M.N.* One feature of the constructive solutions of the lattice girder.International Journal for Computational Civil and Structural Engineering. 2018. 14(4). Pp. 90-97. doi: 10.22337/2587-9618-2018-14-4-90-97

43. *Kirsanov M.N., Kollerova S.E.* Elastic characteristics of a stochastically inhomogeneous medium modeled by a rod grid. Science Almanac. 2017. 3-3(29). Pp. 258-261. doi: 10.17117/na.2017.03.03.258

44. *Kirsanov M.N., Lafickova M.G., Nikitina A.S.* Analysis of the deflection of flat roof trusses. Science Almanac. 2017.3-3 (29). Pp. 262-265. doi: 10.17117/na. 2017.03.03. 262

45. *Kirsanov M.N., Lafickova M.G., Nikitina A.S.* An inductive derivation of the dependence of the arched truss deflection on the number of panels. Science Almanac. 2017. No. 4-3(30). Pp. 205-208. doi: 10.17117/na.2017.04.03.205

46. *Kirsanov M.N., Shi Kaifei* The calculation of the deflectionof the truss with cross bars in the system Maple. Science Almanac. 2017. 3-3(29). Pp. 266-269. doi: 10.17117/na.2017.03.03.266

47. *Kirsanov M.N., Shi Kaifei* The formulas for calculating girders with asymmetric triangular lattice. Science Almanac. 2017. 3-3 (29). Pp. 270-273. 25.3.17 doi: 10.17117/na.2017.03.03.270

48. *Kirsanov M.N., Razananairina P.C.* The formula for deflection of truss with cases of kinematic variability. Postulat. 2017. 9(23).

49. *Kirsanov M.N., Astahov S.V.* The mathematical model of dome covering industrial facility. Architecture and Engineering. 2017. No. 2(2). Pp. 8-14. doi: 10. 23968/2500-0055-2017-2-2-8-14

50. *Kirsanov M.N., Orlov I.V.* The dependence of the deflection of the rod of a statically exterior indeterminate truss on the number of panels. Postulat. 2017. 12(26).

51. *Kirsanov M.N.* An inductive method of calculation of the deflection of the truss regular type. Architecture and Engineering. 2016. No. 3. Vol. 1. Pp. 14-17.

52. *Kirsanov M.N.* Analytical calculation, marginal and comparative analysis of a flat girder. Scientific Herald of the Voronezh State University of Architecture and Civil Engineering. Construction and Architecture. 2016. No. 1 (29). Pp. 84-105.

53. *Kitaev S.S.* Derivation of the formula for the deflection of a cantilevered truss with a rectangular diagonal grid in the computer mathematics system Maple. Postulat. 2018. No. 5-1(31).

54. *Kiyko L.K.* Analytical evaluation of deflection of arched trusses under wind load. Scientific Bulletin. 2016. No. 1(7). Pp. 247-254.

55. *Ponamareva M.A.* The displacement of the support trusses with parallel belts under uniform load. Science Almanac. 2016. No. 4-3(18). Pp. 257-259. doi: 10.17117/na.2016.04.03.257

56. *Rakhmatulina A.R., Smirnova A.A.* The dependence of the deflection of the arched truss loaded on the upper belt, on the number of panels. Science Almanac. 2017. No. 2-3(28). Pp. 268-271.

57. *Rakhmatulina A.R., Smirnova A.A.* The formula for the deflection of a truss loaded at half-span by a uniform load. Postulat. 2018. 3(29).

58. *Rakhmatulina A.R., Smirnova A.A.* Two-parameter derivation of the formula for deflection of the console truss. Postulat. 2018. 5-1(31).

59. *Rakhmatulina A.R., Smirnova A.A.* Analytical calculation and analysis of planar springel truss. Structural mechanics and structures. 2018. No. 2(17). Pp. 72-79.

60. *Ritter A.* Elementare Theorie Und Berechnung Eiserner Dach-Und Brucken-Constructionen. Baümgartner's Buchhandlung.— Leipzig, 1882.

61. *Salimov M.S.* The formula for deflection of a composite truss, loaded on the bottom flange. Science Almanac. 2017. No. 2-3(28). Pp. 272-274. doi: 10.17117/na.2017.02.03.272

62. *Shipaeva A.S.* Calculation of the deflection of girder beam loaded on the bottom flange in the system Maple. Science Almanac. 2016. No. 5-3(19). Pp. 236-239.

63. *Smirnova A.A., Rakhmatulina A.R.* Analytical calculation of the displacement of the truss support. Science Almanac. 2017. No. 2-3(28). Pp. 275-278.

64. *Tinkov D.V., Safonov A.A.* Design Optimization of Truss Bridge Structures of Composite Materials. Journal of Machinery Manufacture and Reliability. 2017. Vol. 46. No. 1. Pp. 46–52. doi: 10.3103/S1052618817010149

65. *Vasilchenko D.I.* Analytical evaluation of displacement of the movable support in the truss with reinforced cross bars. Science Almanac. 2017. No. 4-3(30). Pp. 198-201.

66. *Voropay R.A.* Analytical calculation of the deflection of a beam truss with parallel belts. Postulat. 2018. 6(32). 96.

67. *Voropay R.A.* Derivation of the formula for the deflection of the truss with additional horizontal struts. Postulat. 2018. 6(32). 105.

68. *Voropay R.A.* The derivation of the dependence of the deflection of a beam truss with lateral struts on the number of panels. Postulat. 2018. 6(32). 125.

69. *Voropay R.A., Domanov E.V.* The formula for the dependence of the deflection of a truss with an asymmetric lattice on the number of panels. Postulat. 2018. 6(32). 61.

70. *Voropay R.A., Domanov E.V.* Analytical solution of the problem of shifting a movable support of a truss of arch type in the Maple system. Postulat. 2019. 1.

71. *Voropay R.A., Domanov E.V.* The dependence of the deflection of a planar beam truss with a complex lattice on the number of panels in the system Maple. Postulat. 2019. 1-1(39). 12.

72. *Voropai R.A., Domanov E.V., Kirsanov M.N.* Analysis of the deflection of planar truss with double triangular lattice. Science Almanac. 2016. No. 12-2(26). Pp. 208-211.

73. *Voropai R.A., Kirsanov M.N.* The formula for deflection of flat truss under the action of concentrated load. Science Almanac. 2016. No. 11-2(25). Pp. 305-307. doi: 10.17117/na.2016.11.02.305

74. *Voropai R.A., Kazmiruk I.Yu.* Analytical study of the horizontal stiffness of the flat statically determinate arch truss. Science Almanac. 2016. No. 2-1(6). Pp. 10-12

75. *Voropai R.A., Kirsanov M.N.* On the deformation of spatial cantilever trusses under the action of lateral loads. Science Almanac. 2016. No. 9-2(23). Pp. 17–20. doi: 10.17117/na.2016.09.02.017

76. https://en.wikipedia.org/wiki/Maple_(software)